Clinical Applications of PCR

METHODS IN MOLECULAR BIOLOGY™

John M. Walker, SERIES EDITOR

METHODS IN MOLECULAR BIOLOGY™

Clinical Applications of PCR

Second Edition

Edited by

Y. M. Dennis Lo
Rossa W. K. Chiu
K. C. Allen Chan

Department of Chemical Pathology
The Chinese University of Hong Kong
Hong Kong SAR

HUMANA PRESS ✳ TOTOWA, NEW JERSEY

© 2006 Humana Press Inc.
999 Riverview Drive, Suite 208
Totowa, New Jersey 07512

www.humanapress.com

This publication is printed on acid-free paper. ∞
ANSI Z39.48-1984 (American Standards Institute)

Permanence of Paper for Printed Library Materials.

Production Editor: Jennifer Hackworth

Cover illustration: Sunny Wong

Cover design by Patricia F. Cleary

For additional copies, pricing for bulk purchases, and/or information about other Humana titles, contact Humana at the above address or at any of the following numbers: Tel.: 973-256-1699; Fax: 973-256-8341; E-mail: orders@humanapr.com; or visit our Website: www.humanapress.com

Printed in the United States of America. 10 9 8 7 6 5 4 3 2 1
1-59745-074-X (e-book)
ISSN 1064-3745

Library of Congress Cataloging-in-Publication Data
Clinical applications of PCR / edited by Y.M. Dennis Lo, Rossa W.K.
Chiu, K.C. Allen Chan.-- 2nd ed.
    p. ; cm. -- (Methods in molecular medicine ; 336)
  Includes bibliographical references and index.
  ISBN 1-58829-348-3 (alk. paper)
  1. Polymerase chain reaction--Diagnostic use.
  [DNLM: 1. Polymerase Chain Reaction--methods. QU 450 C241 2006]
I. Lo, Y. M. Dennis. II. Chiu, Rossa W. K. III. Chan, K. C. Allen
IV. Series.
  RB43.8.P64C55 2006
  616.07'56--dc22   2005022727

Preface

Since the invention of the polymerase chain reaction (PCR) in the early 1980s, this technique has rapidly become an indispensable part of modern molecular diagnostics. Without this powerful technology, many of the important developments in modern sciences, including the Human Genome Project, would probably have progressed much more slowly. In the area of molecular diagnostics, PCR has allowed target detection to be performed with unprecedented sensitivity and ease.

It has been several years since the first edition of *Clinical Applications of PCR* was published. During these few years, it is amazing how rapidly technological advances in PCR-based technologies have developed. Important technological advances, notably real-time PCR and mass spectrometry, have revolutionized the field. In particular, real-time PCR has allowed the technique to be performed with improved sensitivity, robustness, and resilience to carryover contamination, as well as in a quantitative manner. These technological developments, together with the indispensable nature of PCR in molecular laboratories everywhere, have led to a vast expansion in the number of clinical applications of PCR.

In the second edition of *Clinical Applications of PCR,* we hope to share with readers the exciting applications of some of these innovations, including PCR for gene expression, methylation, trace molecule, gene dosage, and single cell analysis. It is hoped that the step-by-step protocols and the explanatory notes will help readers to harness the power of these techniques in their laboratories.

Y. M. Dennis Lo
Rossa W. K. Chiu
K. C. Allen Chan

Contents

Contributors

K. C. ALLEN CHAN • *Department of Chemical Pathology, Prince of Wales Hospital, The Chinese University of Hong Kong, Hong Kong SAR*

TIAN-JIAN CHEN • *Department of Medical Genetics, University of South Alabama, Mobile, AL*

STEPHEN S. C. CHIM • *Department of Obstetrics and Gynaecology, Prince of Wales Hospital, The Chinese University of Hong Kong, Hong Kong SAR*

ROSSA W. K. CHIU • *Department of Chemical Pathology, Prince of Wales Hospital, The Chinese University of Hong Kong, Hong Kong SAR*

BRYAN R. COBB • *Institute for Molecular and Human Genetics, Georgetown University Medical Center, Washington, DC*

CHUNMING DING • *Centre for Emerging Infectious Diseases, Prince of Wales Hospital, The Chinese University of Hong Kong, Shatin, New Territories, Hong Kong SAR*

SHIRA GAL • *Nuffield Department of Clinical Laboratory Sciences, University of Oxford, John Radcliffe Hospital, Oxford, UK*

SINUHE HAHN • *Laboratory for Prenatal Medicine, University Women's Hospital, Basel, Switzerland*

WOLFGANG HOLZGREVE • *Laboratory for Prenatal Medicine, University Women's Hospital, Basel, Switzerland*

T. VAUVERT HVIID • *Department of Clinical Biochemistry, H:S Rigshospitalet, Copenhagen University Hospital, Copenhagen, Denmark*

CHING-WAN LAM • *Department of Chemical Pathology, Prince of Wales Hospital, The Chinese University of Hong Kong, Hong Kong SAR*

LISA LEVETT • *Cytogenetic DNA Services Ltd., London, UK*

ANATOLY V. LICHTENSTEIN • *Cancer Research Center, Moscow, Russia*

Y. M. DENNIS LO • *Department of Chemical Pathology, Prince of Wales Hospital, The Chinese University of Hong Kong, Hong Kong SAR*

HOVSEP S. MELKONYAN • *Xenomics Inc., New York, NY/Princeton, NJ*

ENDERS K. O. NG • *Department of Health, Centre for Health Protection, Government of Hong Kong Special Administrative Region, Hong Kong SAR*

ROBERT J. PRYOR • *Department of Pathology, University of Utah School of Medicine, Salt Lake City, UT*

L. DAVID TOMEI • *Xenomics Inc., New York, NY/Princeton, NJ*

ix

NANCY B. Y. TSUI • *Department of Chemical Pathology, Prince of Wales Hospital, The Chinese University of Hong Kong, Hong Kong SAR*

SAMUIL R. UMANSKY • *Xenomics Inc., New York, NY/Princeton, NJ*

JAMES S. WAINSCOAT • *Leukaemia Research Fund Molecular Haematology Unit, Nuffield Department of Clinical Laboratory Sciences, John Radcliffe Hospital, University of Oxford, Oxford, UK*

CARL T. WITTWER • *Department of Pathology, University of Utah School of Medicine, Salt Lake City, UT*

IVY H. N. WONG • *Department of Obstetrics and Gynaecology, Prince of Wales Hospital, The Chinese University of Hong Kong, Hong Kong SAR*

LEE-JUN C. WONG • *Department of Molecular and Human Genetics, Baylor College of Medicine, Houston, TX*

BERNHARD ZIMMERMANN • *Centre for Research in Biomedicine, University of the West of England, Bristol, UK*

1

Introduction to the Polymerase Chain Reaction

Y. M. Dennis Lo and K. C. Allen Chan

Summary

The polymerase chain reaction (PCR) is an in vitro method for the amplification of DNA. Since the introduction of the PCR in 1985, it has become an indispensable technique for many applications in scientific research and clinical and forensic investigations. In this chapter, the principle and setup of PCR, as well as the methods for analyzing PCR products, will be discussed.

Key Words: Polymerase chain reaction; PCR; principle.

1. Introduction

The polymerase chain reaction (PCR) is an in vitro method for the amplification of DNA that was introduced in 1985 *(1)*. The principle of the PCR is elegantly simple, but the resulting method is extremely powerful. The adoption of the thermostable *Taq* polymerase in 1988 greatly simplified the process and enabled the automation of PCR *(2)*. Since then, a large number of applications that are based on the basic PCR theme have been developed. The versatility and speed of PCR have revolutionized molecular diagnostics, allowing the realization of a number of applications that were impossible in the pre-PCR era. This chapter offers an introductory guide to the process.

2. Principle of the PCR

PCR may be regarded as a simplified version of the DNA replication process that occurs during cell division. Basic PCR consists of three steps: thermal denaturation of the target DNA, primer annealing of synthetic oligonucleotide primers, and extension of the annealed primers by a DNA polymerase (**Fig. 1**). This three-step cycle is then repeated a number of times, each time approximately doubling the number of product molecules. The amplification factor is

From: *Methods in Molecular Biology, vol. 336: Clinical Applications of PCR*
Edited by: Y. M. D. Lo, R. W. K. Chiu, and K. C. A. Chan © Humana Press Inc., Totowa, NJ

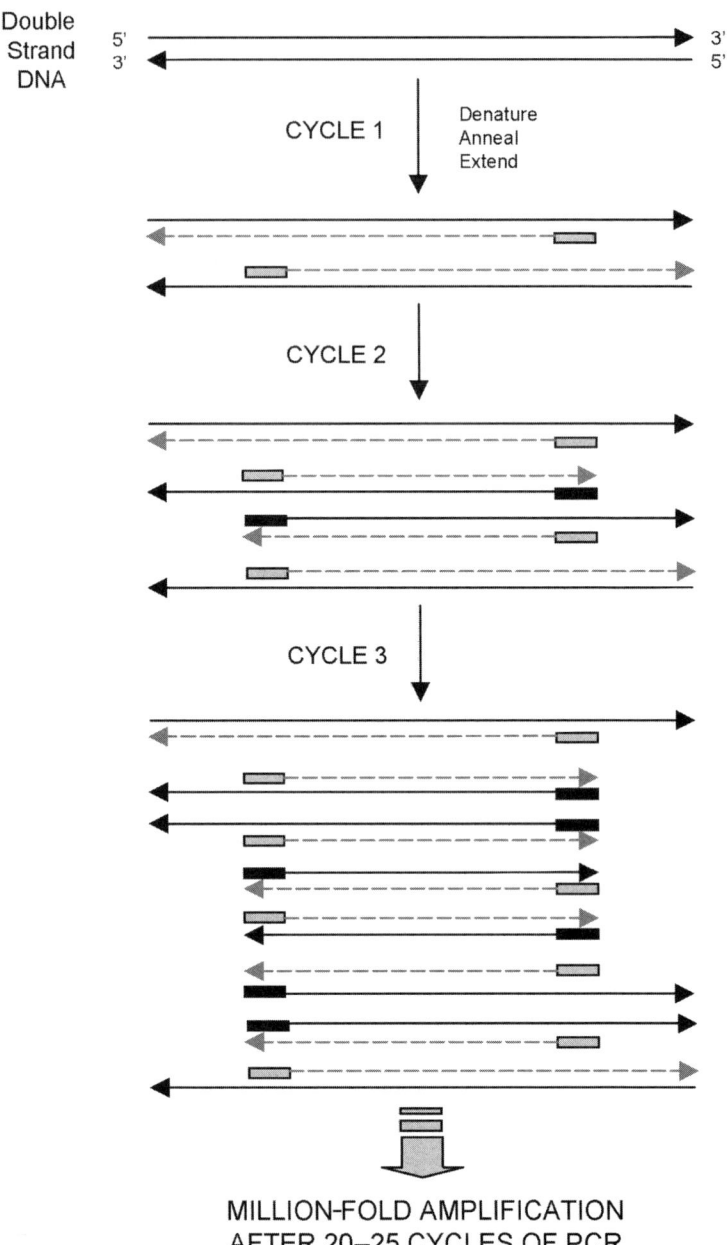

MILLION-FOLD AMPLIFICATION
AFTER 20–25 CYCLES OF PCR

Fig. 1. Schematic representation of the polymerase chain reaction. The newly synthesized DNA is indicated by dotted lines in each cycle. Oligonucleotide primers are indicated by solid rectangles. Each DNA strand is marked with an arrow indicating the 5' to 3' orientation.

given by the equation $x(1 + E)^n$, where x = initial amount of target, E = efficiency of amplification, and n = number of PCR cycles. After a few cycles, the resulting product is of the size determined by the distance between the 5' ends of the two primers. With the performance of a previous reverse transcription step, PCR can also be applied to RNA.

3. Composition of the PCR

Conventional PCR is usually performed in a volume of 10–100 µL. Deoxynucleoside triphosphates (dATP, dCTP, dGTP, and dTTP) at a concentration of 200 µ*M* each, 10 to 100 pmol of each primer, the appropriate salts, buffers, and DNA polymerase are included. Many manufacturers have included reaction buffer with their DNA polymerase, and this proves convenient for newcomers to the PCR process. With the development of nanotechnology, there is a trend for the miniaturization of the PCR reaction. PCR reaction has been reported to have been successfully performed in an 86-pL microchamber *(3)*.

4. Primers

Primers are designed to flank the sequence of interest. Oligonucleotide primers are usually between 18 and 30 bases long, with a GC content of about 50%. Complementarity at the 3' ends of the primers should be avoided in order to decrease the likelihood of forming the primer-dimer artifact. Runs of three or more Cs or Gs at the 3' ends of the primers should be avoided in order to decrease the probability of priming GC-rich sequences nonspecifically. A number of computer programs that assist primer design are available. However, for most applications, PCR is sufficiently forgiving in that most primer pairs seem to work. The primers are generally positioned between 100 and 1000 bp apart. It should be noted, however, that for high-sensitivity applications, shorter PCR products are preferred. For most applications, purification of the PCR primers is not necessary. To simplify subsequent operations, it is recommended that all primers be diluted to the same concentration (e.g., 50 pmol/µL) so that the same volume of each primer is required for each reaction. Some primer pairs seem to fail without any obvious reason, and when difficulty arises, one simple solution is to change one or both of the primers.

Several sets of primers can be included in a single PCR reaction chamber for the simultaneous amplification of different targets from a common DNA source. In order to achieve similar efficiencies in all PCR reactions, the melting temperatures of different sets of primers should be as close as possible. The melting temperature is defined as the temperature at which one-half of the oligos form a duplex with the complementary sequence. There are a wide range of applications for this multiplex PCR technique, e.g., detection of infectious agents *(4)* and oncogene mutations *(5)*.

5. Steps of the PCR
5.1. Thermal Denaturation

A common cause of failed PCR is inadequate denaturation of the DNA target. We typically use an initial denaturation temperature of 94°C for 8 min. Modified *Taq* polymerase, e.g., Ampli*Taq* Gold®, can be activated during this initial denaturing process to achieve a Hot Start PCR. For subsequent cycles, 94°C for 1–2 min is usually adequate. As the targets of later PCR cycles are mainly PCR products rather than genomic DNA, it has been suggested that the denaturation temperature may be lowered after the first 10 cycles so as to avoid excessive thermal denaturation of the *Taq* polymerase *(6)*. The half-life of *Taq* DNA polymerase activity is more than 2 h at 92.5°C, 40 min at 95°C, and 5 min at 97.5°C.

5.2. Primer Annealing

The temperature and length of time required for primer annealing depends on the base composition and the length and concentration of the primers. Using primers of 18–30 bases long with approx 50% GC content and an annealing step of 55°C for 1–2 min is a good start. In certain primer–template pairs, a difference in the annealing temperature as small as 1–2°C will make the difference between specific and nonspecific amplification. If the annealing temperature is >60°C, it is possible to combine the annealing and extension step together into a two-step PCR cycle.

5.3. Primer Extension

Primer extension is typically carried out at 72°C, which is close to the optimum temperature of the *Taq* polymerase. An extension time of 1 min is generally enough for products up to 2 kb in length. Longer extension times (e.g., 3 min) may be helpful in the first few cycles for amplifying a low copy number target, or at later cycles when product concentration exceeds enzyme concentration.

6. Cycle Number

The number of cycles needed is dependent on the copy number of the target. As a rule of thumb, amplifying 10^5 template molecules to a signal visible on an ethidium bromide-stained agarose gel requires 25 cycles. Assuming that we use 1 min each for denaturation, annealing, and extension, the whole process can be completed in approx 2–3 h (with extra time allowed for the lag phase taken by the heat block to reach a certain temperature). Similarly, 10^4, 10^3, and 10^2 target molecules will require 30, 35, and 40 cycles, respectively.

7. PCR Plateau

There is a limit to how many product molecules a given PCR can produce. For a 100-µL PCR, the plateau is about 3–5 pmol *(7)*. The plateau effect is caused by the accumulation of product molecules, which results in a significant degree of annealing between complementary product strands rather than between the primers and templates. Furthermore, the finite amount of enzyme molecules present will be unable to extend all the primer–template complexes in the given extension time.

8. Sensitivity

The sensitivity of PCR is related to the number of target molecules, the complexity of nontarget molecules, and the number of PCR cycles. Generally, PCR is capable of amplification from a single target molecule *(8)*. This single-molecule capability has allowed the development of single-molecule genotyping *(8,9)* and pre-implantation diagnosis *(10)*. In these applications, the single target molecule is essentially bathed in PCR buffer—in other words, in a low-complexity environment. In situations in which the complexity of the environment is high, the reliability of single-molecule PCR decreases and strategies such as nesting and Hot Start PCR *(11,12)* are necessary for achieving maximum sensitivity. The sensitivity of PCR has also allowed it to be used in situations in which the starting materials have been partially degraded, e.g., with formalin-fixed, paraffin wax-embedded materials.

9. PCR Fidelity

The fidelity of amplification by PCR is dependent on several factors: annealing/extension time, annealing temperature, dNTP concentration, $MgCl_2$ concentration, and the type of DNA polymerase used. In general, the rate of misincorporation may be reduced by minimizing the annealing/extension time, maximizing the annealing temperature, and minimizing the dNTP and $MgCl_2$ concentration *(13)*. Eckert and Kunkel reported an error rate per nucleotide polymerized at 70°C of 10^{-5} for base substitution and 10^{-6} for frameshift errors under optimized conditions *(13)*. The use of a DNA polymerase with proofreading activity reduces the rate of misincorporation. For example, the DNA polymerase from *Thermococcus litoralis*, which has proofreading activity, misincorporates at 25% of the rate of the *Taq* polymerase, which lacks such activity *(14)*. Interestingly, the combination of enzymes with and without proofreading activity has enabled the amplification of extremely long PCR products.

For most applications, product molecules from individual PCRs are analyzed as a whole population, and rare misincorporated nucleotides in a small proportion of molecules pose little danger to the accurate interpretation of data. However, for sequence analysis of cloned PCR products, errors owing to misincorporation may sometimes complicate data interpretation. Thus, it is advisable to analyze multiple clones from a single PCR or to clone PCR products from several independent amplifications. Another application in which misincorporation may result in error of interpretation is in the amplification of low-copy-number targets (e.g., single-molecule PCR). In these situations, if a misincorporation happens in an early PCR cycle (the extreme case being in cycle 1), the error will be passed on to a significant proportion of the final PCR products. Hence, in these applications, the amplification conditions should be carefully optimized.

10. PCR Thermocyclers

One of the main attractions of PCR is its ability to be automated. A number of thermocyclers are available from different manufacturers. These thermocyclers differ in the design of the cooling systems, tube capacity, number of heating blocks, program memory, and thermal uniformity. Units with multiple heating blocks are very valuable for arriving at the optimal cycling profile for a new set of primers, as multiple conditions can be tested simultaneously. Tube capacity generally ranges from 32 to 384 wells and should be chosen with the throughput of the laboratory in mind. Most thermocyclers nowadays have heated covers, making the addition of oil to the PCR mixture obsolete. A variant of thermocycler using the convection of PCR mixtures between the annealing and denaturing temperature has been developed *(15,16)*. This convection-driven PCR can significantly shorten the time for a PCR, and the exponential amplification can reach 100,000-fold within 25 min *(17)*.

11. Analysis and Processing of PCR Product

The amplification factor produced by PCR simplifies the analysis and detection of the amplification products. In general, analytical methods for conventional DNA sources are also applicable to PCR products.

11.1. Electrophoresis

Agarose gel electrophoresis followed by ethidium bromide staining represents the most common way to analyze PCR products. A 1.5% agarose gel is adequate for the analysis of PCR products from 150 to 1000 bp. DNA markers of different size ranges are available commercially. The development of capillary electrophoresis has significantly increased the throughput and resolution of PCR product analysis. The improved resolution of capillary electrophoresis has allowed the discrimination of a single nucleotide difference in size.

11.2. Restriction of PCR Products

Restriction mapping is a common way to verify the identity of a PCR product. It is also a simple method of detecting restriction site polymorphisms and of detecting mutations that are associated with the creation or destruction of restriction sites. There is no need to purify the PCR product prior to restriction, and most restriction enzymes are functional in a restriction mix in which the PCR product constitutes up to one-half of the total volume.

11.3. Sequence-Specific Oligonucleotide Hybridization

Sequence-specific oligonucleotide hybridization is a powerful method for detecting the presence of sequence polymorphisms in a region amplified by PCR. Short oligonucleotides are synthesized and labeled (either radioactively or nonradioactively), allowed to hybridize to dot blots of the PCR products *(8)*, and washed under conditions that allow the discrimination of a single nucleotide mismatch between the probe and the target PCR product.

For the detection of a range of DNA polymorphisms at a given locus, the hybridization can be performed "in reverse," that is, with the oligonucleotides immobilized onto the filter. Labeled amplified products from target DNA are then hybridized to the filters and washed under appropriate conditions *(18)*. When the oligonucleotides are immobilized onto a glass slide, it can be miniaturized to form a microarray. Labeled amplified products from DNA can be hybridized to the microarray for the investigation of chromosomal aberrations *(19–21)*.

11.4. Cloning of PCR Product

PCR products may be cloned easily using conventional recombinant DNA technology. To facilitate cloning of PCR products into vectors, restriction sites may be incorporated into the primer sequences. Digestion of the PCR products with the appropriate restriction enzymes will then allow "sticky end" ligation into similarly restricted vector DNA.

11.5. Real-Time PCR

A fluorescent probe hybridizing to a region between the 3' ends of the primers can be incorporated into the PCR reaction *(22)*. In a particular incarnation of this technique, during each round of PCR, the fluorescent probes would be cleaved and the detectable fluorescent signal would increase. The amount of increased fluorescent signal is proportional to the number of newly synthesized amplicons. A major advantage of real-time PCR is that it can be used to determine the amount of initial templates.

11.6. Mass Spectrometry

The PCR products can be analyzed by matrix-assisted laser desorption/ionization time-of-flight mass spectrometry. The PCR products are subjected to a separate primer extension reaction. The extension products are dotted on a chip and ionized by laser. The molecular mass of the extension products can then be determined by the time required to reach the detector. This detection method has been applied for high-throughput single nucleotide polymorphism genotyping *(23–25)*, mutation detection *(26)*, and prenatal diagnosis *(27)*.

12. Conclusion

The versatility of PCR has made it one of the most widely used methods in molecular diagnosis. The number of PCR-based applications has continued to increase rapidly and has impacted oncology (*see* Chapters 4, 10, and 13), genetics (*see* Chapters 8 and 12), and pathogen detection (*see* Chapters 14–16). In *Clincial Applications of PCR: Second Edition*, we attempt to present some of the most important clinical applications of PCR.

References

1. Saiki, R. K., Scharf, S., Faloona, F., et al. (1985) Enzymatic amplification of beta-globin genomic sequences and restriction site analysis for diagnosis of sickle cell anemia. *Science* **230,** 1350–1354.
2. Saiki, R. K., Gelfand, D. H., Stoffel, S., et al. (1988) Primer-directed enzymatic amplification of DNA with a thermostable DNA polymerase. *Science* **239,** 487–491.
3. Nagai, H., Murakami, Y., Morita, Y., Yokoyama, K., and Tamiya, E. (2001) Development of a microchamber array for picoliter PCR. *Anal. Chem.* **73,** 1043–1047.
4. Puppe, W., Weigl, J. A., Aron, G., et al. (2004) Evaluation of a multiplex reverse transcriptase PCR ELISA for the detection of nine respiratory tract pathogens. *J. Clin. Virol.* **30,** 165–174.
5. Tournier, I., Paillerets, B. B., Sobol, H., et al. (2004) Significant contribution of germline BRCA2 rearrangements in male breast cancer families. *Cancer Res.* **64,** 8143–8147.
6. Yap, E. P. and McGee, J. O. (1991) Short PCR product yields improved by lower denaturation temperatures. *Nucleic Acids Res.* **19,** 1713.
7. Higuchi, R., Krummel, B., and Saiki, R. K. (1988) A general method of in vitro preparation and specific mutagenesis of DNA fragments: study of protein and DNA interactions. *Nucleic Acids Res.* **16,** 7351–7367.
8. Li, H. H., Gyllensten, U. B., Cui, X. F., Saiki, R. K., Erlich, H. A., and Arnheim, N. (1988) Amplification and analysis of DNA sequences in single human sperm and diploid cells. *Nature* **335,** 414–417.
9. Jeffreys, A. J., Kauppi, L., and Neumann, R. (2001) Intensely punctate meiotic recombination in the class II region of the major histocompatibility complex. *Nat. Genet.* **29,** 217–222.

10. Moutou, C., Gardes, N., and Viville, S. (2004) New tools for preimplantation genetic diagnosis of Huntington's disease and their clinical applications. *Eur. J. Hum. Genet.* **12**, 1007–1014.
11. Chou, Q., Russell, M., Birch, D. E., Raymond, J., and Bloch, W. (1992) Prevention of pre-PCR mis-priming and primer dimerization improves low-copy-number amplifications. *Nucleic Acids Res.* **20**, 1717–1723.
12. Birch, D. E. (1996) Simplified hot start PCR. *Nature* 381, 445–446.
13. Eckert, K. A. and Kunkel, T. A. (1991) DNA polymerase fidelity and the polymerase chain reaction. *PCR Methods Appl.* **1**, 17–24.
14. Cariello, N. F., Swenberg, J. A., and Skopek, T. R. (1991) Fidelity of Thermococcus litoralis DNA polymerase (Vent) in PCR determined by denaturing gradient gel electrophoresis. *Nucleic Acids Res.* **19**, 4193–4198.
15. Krishnan, M., Ugaz, V. M., and Burns, M. A. (2002) PCR in a Rayleigh-Benard convection cell. *Science* **298**, 793.
16. Wheeler, E. K., Benett, W., Stratton, P., et al. (2004) Convectively driven polymerase chain reaction thermal cycler. *Anal. Chem.* **76**, 4011–4016.
17. Braun, D., Goddard, N. L., and Libchaber, A. (2003) Exponential DNA replication by laminar convection. *Phys. Rev. Lett.* **91**, 158103.
18. Saiki, R. K., Walsh, P. S., Levenson, C. H., and Erlich, H. A. (1989) Genetic analysis of amplified DNA with immobilized sequence-specific oligonucleotide probes. *Proc. Natl. Acad. Sci. USA* **86**, 6230–6234.
19. Dhami, P., Coffey, A. J., Abbs, S., et al. (2005) Exon array CGH: detection of copy-number changes at the resolution of individual exons in the human genome. *Am. J. Hum. Genet.* **76**, 750–762.
20. Devries, S., Nyante, S., Korkola, J., et al. (2005) Array-based comparative genomic hybridization from formalin-fixed, paraffin-embedded breast tumors. *J. Mol. Diagn.* **7**, 65–71.
21. Hu, D. G., Webb, G., and Hussey, N. (2004) Aneuploidy detection in single cells using DNA array-based comparative genomic hybridization. *Mol. Hum. Reprod.* **10**, 283–289.
22. Heid, C. A., Stevens, J., Livak, K. J., and Williams, P. M. (1996) Real time quantitative PCR. *Genome Res.* **6**, 986–994.
23. Nelson, M. R., Marnellos, G., Kammerer, S., et al. (2004) Large-scale validation of single nucleotide polymorphisms in gene regions. *Genome Res.* **14**, 1664–1668.
24. Buetow, K. H., Edmonson, M., MacDonald, R., et al. (2001) High-throughput development and characterization of a genomewide collection of gene-based single nucleotide polymorphism markers by chip-based matrix-assisted laser desorption/ionization time-of-flight mass spectrometry. *Proc. Natl. Acad. Sci. USA* **98**, 581–584.
25. Mohlke, K. L., Erdos, M. R., Scott, L. J., et al. (2002) High-throughput screening for evidence of association by using mass spectrometry genotyping on DNA pools. *Proc. Natl. Acad. Sci. USA* **99**, 16,928–16,933.
26. Lleonart, M. E., Ramon y Cajal, S., Groopman, J. D., and Friesen, M. D. (2004) Sensitive and specific detection of K-ras mutations in colon tumors by short oligonucleotide mass analysis. *Nucleic Acids Res.* **32**, e53.

27. Ding, C., Chiu, R. W. K., Lau, T. K., et al. (2004) MS analysis of single-nucleotide differences in circulating nucleic acids: Application to noninvasive prenatal diagnosis. *Proc. Natl. Acad. Sci. USA* **101,** 10,762–10,767.

2

Setting Up a Polymerase Chain Reaction Laboratory

Y. M. Dennis Lo and K. C. Allen Chan

Summary

One of the most important attributes of the polymerase chain reaction (PCR) is its exquisite sensitivity. However, the high sensitivity of PCR also renders it prone to false-positive results because of, for example, exogenous contamination. Good laboratory practice and specific anti-contamination strategies are essential to minimize the chance of contamination. Some of these strategies, for example, physical separation of the areas for the handling samples and PCR products, may need to be taken into consideration during the establishment of a laboratory. In this chapter, different strategies for the detection, avoidance, and elimination of PCR contamination will be discussed.

Key Words: False-positive PCR; anti-contamination strategies.

1. Introduction

One of the most important attributes of the polymerase chain reaction (PCR) is its exquisite sensitivity. However, this high sensitivity has also given PCR its main weakness, namely, its tendency to produce false-positive results owing to exogenous contamination *(1,2)*. Contamination avoidance is therefore the single most important consideration when setting up a PCR laboratory *(3)*, especially one designed to generate diagnostic information *(4–7)*. In many situations, precautions that are normally taken in the handling of microbiological materials are equally applicable to PCR-related procedures *(7)*.

2. Sources of Contamination

There are four main sources of PCR contamination. The most important one is PCR products from previous amplifications, the so-called carryover contamination *(3)*. Because of the enormous amplification power of PCR and its ability to generate up to 10^{12} product molecules in a single reaction, this is the

From: *Methods in Molecular Biology, vol. 336: Clinical Applications of PCR*
Edited by: Y. M. D. Lo, R. W. K. Chiu, and K. C. A. Chan © Humana Press Inc., Totowa, NJ

most serious source of contamination. When such large amounts of PCR products are generated repeatedly over a period of time, the potential for contamination becomes increasingly high. This is further compounded by the fact that many diagnostic applications require PCR to perform at its highest sensitivity, namely, at the single-molecule level. Under these circumstances, even one of the billions of molecules generated from a single reaction is enough to generate a false-positive result. The second source of contamination is cloned DNA previously handled in the laboratory. The third type is sample-to-sample contamination. This source of contamination is most detrimental to samples that require extensive processing prior to amplification. The fourth source is the ubiquitously present template DNA in the environment from the laboratory personnel and reagents used for DNA extraction and PCR *(8–10)*.

3. Principles of Contamination Avoidance

Like many problems, avoidance is better than cure, and PCR contamination is no exception. The main principles of contamination avoidance in PCR are:

1. Strict physical separation of individual PCR-related maneuvers: we recommend the use of three distinct areas for the sample preparation stage, the PCR setup stage, and the post-PCR stage. This applies as much to the performance of laboratory procedure as to equipment. Thus, every piece of equipment, no matter how small, should be restricted to each area. This applies to laboratory notebooks, which should not be carried between different areas. If transfer of items is essential, then the direction should be from the pre-PCR area to the post-PCR area and never the reverse.

    a. Sample preparation area: this area is for the processing of sample materials, such as the extraction of DNA and RNA. No PCR products should ever be allowed into the area. Dedicated equipment and reagents should be reserved solely for sample preparation purposes, including pipetting equipment and laboratory coats. Gloves should be worn at all times and changed frequently. In general, the simpler the sample processing is, the less chance there is of introducing contamination. Dedicated storage facilities, e.g., freezers, should be available for sample preparation alone.

    b. PCR setup area: it is recommended that the setting up of PCR reactions be performed in a laminar flow hood. The defined area of the hood facilitates the maintenance of cleanliness of the area. Dedicated equipment and storage facilities should be available near the PCR setup area. A separate area should be available for the addition of samples to the PCR reagents. DNA or RNA samples should never be allowed inside the PCR setup hood.

    c. PCR machine: the location of the PCR machine depends on the exact amplification requirements. For PCR applications involving a single round of PCR and in which it is not required that individual PCR tubes be opened for the addition or sampling of reagents prior to analysis, the PCR machine may be located in the post-PCR area (*see* **item d**). However, for applications in which the PCR tubes must be opened, e.g., for nested PCR, the PCR machine should be located at a fourth isolated area separated from sample preparation, PCR setup, and the post-PCR areas. In nested PCR, a dedicated set of pipets should be allocated for this purpose.

d. Post-PCR area: this is the area reserved for the analysis of PCR products, including electrophoresis, restriction analysis, and mass spectrometry. No items from the post-PCR area should be allowed back into the aforementioned areas. It is important to note that this includes items such as notebooks and pens.

2. Laboratory practice designed to minimize the risk of contamination:

a. All PCR reagents should be aliquoted, and reagents that can be autoclaved should be so treated.

b. Use and change gloves frequently. Kitchin et al. have advocated the use of face and head masks, as certain individuals appear more prone to the shedding of contaminants *(11)*.

c. Positive displacement pipets or aerosol-resistant pipets should be used.

d. When multiple reactions are needed, it is helpful to set up a master mix to reduce the number of maneuvers, and thus reduce the chance of possible contamination.

e. The number of PCR cycles should be kept to a minimum, as excessively sensitive assays are more prone to contamination *(12)*.

f. When given a choice, disposable items are preferable to items that must be washed prior to being reused.

g. If possible, different personnel should be allocated to the pre-PCR and post-PCR parts of the project. If this is not practical, then it is preferable to schedule the project or work week such that the pre-PCR and post-PCR procedures are performed on different days.

h. The use of closed PCR systems, e.g., the TaqMan® system *(13)*, which use fluorescence signals for detecting PCR products, can eliminate the opening of the amplification vessels and post-PCR sample handling. Therefore, carryover contamination using these systems is much less of a problem than conventional systems. This is especially important for clinical diagnostic applications *(5)*.

3. Use of specific anti-contamination measures:

a. Ultraviolet (UV) irradiation: Sarkar and Sommer describe the use of UV irradiation to damage any contaminating DNA prior to the addition of DNA template *(14,15)*. As this method relies on the crosslinking of adjacent thymidine residues, the sequence of the PCR target influences the decontaminating efficiency of the method *(16)*. Certain primers appear to be more sensitive to the damaging effect of UV light, and may need to be added after the irradiation step. Furthermore, the hydration status of DNA appears to have a great influence on its susceptibility to UV irradiation in that dry DNA seems much more resistant to the damaging effect of UV *(16,17)*. This latter fact means that there are limitations to the use of UV for sterilizing dry laboratory surfaces. Ultimately, clean laboratory practices and physical separation remain the most important anti-contamination measures, with UV irradiation providing an additional margin of protection.

b. Restriction enzyme treatment: restriction enzymes that cleave within the target sequence for PCR may be used to restrict any contaminating sequence prior to the addition of the target *(18,19)*. Following decontamination, the enzyme is destroyed by thermal denaturation (i.e., 94°C for 10 min; thus, thermostable restriction enzymes such as *Taq*I should not be used for this purpose) before addition of the template DNA. In a model system, Furrer et al. showed that restriction with *Msp*I (10 U for 1 h) reduced contamination by a factor of 5 to 10 without impairing the efficiency of PCR *(19)*.

c. DNase I treatment: this approach is similar to that in **item b** except that DNase I is used. Furrer et al. showed that prior treatment with 0.5 U of DNase I for 30 min reduced contamination by a factor of 1000 without impairing the efficiency of PCR *(19)*.

d. Incorporation of dUTP and treatment with uracil-*N*-glycosylase (UNG): as the carryover of PCR products from previous amplification experiments constitutes a predominant source of PCR contamination, the ability to selectively destroy PCR products, but not template DNA, presents one way to reduce contamination. Such an approach is described by Longo et al., who substituted dUTP for dTTP during PCR *(20)*. Carryover PCR products containing dUs can then be destroyed prior to subsequent amplification experiments by incubation with UNG. It should be noted that when dUTP is used instead of dTTP, the $MgCl_2$ concentration often must be readjusted: typically, dUTP is used at 600 mM with 3 mM $MgCl_2$. Following the initial thermal denaturation, UNG activity is destroyed; thus, the newly synthesized PCR products are not degraded. However, UNG may regain some of its activity when the temperature is below 50°C, and thus an annealing temperature of over 50°C should be used *(21)* and all completed PCR containing UNG should be kept at 72°C until analysis. UNG treatment has been reported to result in a 10^7- *(22)* to 10^9-fold *(23)* reduction in amplicon concentration. In our experience, there is a very slight reduction in sensitivity in PCR systems incorporating dUTP and UNG treatment, although up to a 10-fold reduction in sensitivity has been described *(22)*. This method can also be applied to reverse-transcription (RT)-PCR because PCR products which contain deoxyribose uracil are digested by UNG preferentially to ribose uracil-containing RNA with the optimization of the concentration of UNG and the time and temperature of enzyme digestion *(24,25)*. However, it should be remembered that this method is only effective against dU-containing PCR products. Thus, carryover contamination owing to conventional PCR product lacking dUs cannot be eradicated using this method.

e. Incorporation of isopsoralen compound: Cimino et al. describe adding a photochemical reagent before PCR and activation after the amplification is completed *(26)*. The reagent will then crosslink the two strands of the PCR product and render them unamplifiable. The crosslinking is most effective at 5°C and under UV intensity of more than 27 mW/cm^2 (e.g., in an HRI-300 chamber) *(27)*. This method has been shown to be similar in decontaminating efficiency to the UNG method, and results in the elimination of at least 10^9 copies of contaminating PCR products *(23)*.

f. Exonuclease digestion: it was demonstrated that certain exonucleases, e.g., exonuclease III and T7 exonuclease, when added to fully assembled PCR reactions, were able to render carryover PCR product molecules non-amplifiable but would spare identical target sequences in genomic DNA *(28,29)*. In a model system, a 30-min incubation with exonuclease III was able to degrade 5 × 10^5 copies of carryover amplicons *(28)*. Several mechanisms for the selectivity against PCR products have been postulated: Zhu et al. attributed it to the relatively long chain length of genomic DNA, which might resist degradation by exonucleases better than the comparatively short PCR products *(28)*, and Muralidhar and Steinman, in an ingenious series of experiments, demonstrated that part of this selectivity has a geometric explanation *(29)*. Thus, for any stretch of DNA to be amplifiable by a specific pair of primers following T7 exonuclease treatment, the primer binding sites should be situated on the same side with respect to the geometric center of the molecule. As it is extremely unlikely that a particular genomic target would straddle the center of any stretch of genomic DNA (essentially produced by random shearing during DNA extraction), this form of exonuclease treatment would spare the genomic target. The situation with carryover PCR products, however, is completely different, as the primer binding sites are located at opposite ends of the molecules and thus would span the geometric center of the molecule. Exonuclease treatment for the prevention of PCR carryover, therefore, possesses the chief advantage of uridine incorporation and glycosylation in that the completed reaction tubes do not have to be reopened for the addition of the target and/or *Taq* polymerase. Furthermore, exonuclease treatment has the added advantage of being able to destroy even nonuridine-containing PCR products, and would be very useful in an already contaminated environment.

4. Detection of Contamination

Monitoring for contamination is probably as important as measures to prevent it. It is a reality that contamination will be experienced by most, if not all, workers using PCR. To facilitate the monitoring of contamination, the following measures should be undertaken:

1. Negative controls should be included in every PCR experiment. To detect sporadic contamination, multiple controls are usually required. Different negative controls, testing the different stages in the PCR process at which contamination may occur, should be included. PCR reagent controls will only test for contamination of the reagents, but not the sample preparation stage.

2. In certain applications, PCR products from different samples are expected to have different sequences, e.g., sequence variations in bacteria occurring at different times. In these situations, sequencing of PCR products *(30)* or methods that reflect the sequence variation, e.g., heteroduplex analysis *(31)* and single-strand conformation polymorphism (SSCP) analysis *(32)*, are helpful in verifying the genuineness of a positive result.

5. Remedial Measures

Once contamination has been detected, all diagnostic work should be stopped until the source of contamination has been eliminated. In many situations, discarding all suspected reagents is all that is required to cure the problem. In cases in which the equipment is contaminated, thorough cleansing or even replacing the culprit equipment may be necessary. In serious situations, changing to a new primer set that amplifies a different target segment of DNA may be the only method of solving the problem.

6. Automation

Automated nucleic acid extraction (e.g., MagNA Pure®) and liquid handling systems (e.g., Biomek® FX) are now available for the high-throughput nucleic acid extraction and preparing of PCR mixtures. The yield and contamination rates of the automated nucleic acid extraction methods have been shown to be comparable with the manual methods *(33–35)*. These automated platforms are particularly useful in diagnostic laboratories handling a large amount of samples or samples with high infectious risks.

7. Conclusion

Contamination is the single most important obstacle to using PCR reliably for diagnostic purposes. Contamination can only be avoided by meticulous attention to good laboratory-operating details and the exercise of common sense. When coupled with monitoring systems aimed at detecting contamination, reliable PCR, even at high sensitivity, should be a realizable goal.

References

1. Lopez-Rios, F., Illei, P. B., Rusch, V., and Ladanyi, M. (2004) Evidence against a role for SV40 infection in human mesotheliomas and high risk of false-positive PCR results owing to presence of SV40 sequences in common laboratory plasmids. *Lancet* **364,** 1157–1166.
2. Lo, Y. M. D., Mehal, W. Z., and Fleming, K. A. (1988) False-positive results and the polymerase chain reaction. *Lancet* **2,** 679.
3. Kwok, S. and Higuchi, R. (1989) Avoiding false positives with PCR. *Nature* **339,** 237–238.

4. Millar, B. C., Xu, J., and Moore, J. E. (2002) Risk assessment models and contamination management: implications for broad-range ribosomal DNA PCR as a diagnostic tool in medical bacteriology. *J. Clin. Microbiol.* **40,** 1575–1580.
5. Mackay, I. M. (2004) Real-time PCR in the microbiology laboratory. *Clin. Microbiol. Infect.* **10,** 190–212.
6. Aslanzadeh, J. (2004) Preventing PCR amplification carryover contamination in a clinical laboratory. *Ann. Clin. Lab. Sci.* **34,** 389–396.
7. Borst, A., Box, A. T., and Fluit, A. C. (2004) False-positive results and contamination in nucleic acid amplification assays: suggestions for a prevent and destroy strategy. *Eur. J. Clin. Microbiol. Infect. Dis.* **23,** 289–299.
8. Urban, C., Gruber, F., Kundi, M., Falkner, F. G., Dorner, F., and Hammerle, T. (2000) A systematic and quantitative analysis of PCR template contamination. *J. Forensic Sci.* **45,** 1307–1311.
9. Mohammadi, T., Reesink, H. W., Vandenbroucke-Grauls, C. M., and Savelkoul, P. H. (2005) Removal of contaminating DNA from commercial nucleic acid extraction kit reagents. *J. Microbiol. Methods* **61,** 285–288.
10. Mohammadi, T., Reesink, H. W., Vandenbroucke-Grauls, C. M., and Savelkoul, P. H. (2003) Optimization of real-time PCR assay for rapid and sensitive detection of eubacterial 16S ribosomal DNA in platelet concentrates. *J. Clin. Microbiol.* **41,** 4796–4798.
11. Kitchin, P. A., Szotyori, Z., Fromholc, C., and Almond, N. (1990) Avoidance of PCR false positives. *Nature* **344,** 201.
12. Yang, D. Y., Eng, B., and Saunders, S. R. (2003) Hypersensitive PCR, ancient human mtDNA, and contamination. *Hum. Biol.* **75,** 355–364.
13. Heid, C. A., Stevens, J., Livak, K. J., and Williams, P. M. (1996) Real time quantitative PCR. *Genome Res.* **6,** 986–994.
14. Sarkar, G. and Sommer, S. S. (1993) Removal of DNA contamination in polymerase chain reaction reagents by ultraviolet irradiation. *Methods Enzymol.* **218,** 381–388.
15. Sarkar, G. and Sommer, S. S. (1990) Shedding light on PCR contamination. *Nature* **343,** 27.
16. Sarkar, G. and Sommer, S. S. (1991) Parameters affecting susceptibility of PCR contamination to UV inactivation. *Biotechniques* **10,** 590–594.
17. Fairfax, M. R., Metcalf, M. A., and Cone, R. W. (1991) Slow inactivation of dry PCR templates by UV light. *PCR Methods Appl.* **1,** 142–143.
18. Handyside, A. H., Pattinson, J. K., Penketh, R. J., Delhanty, J. D., Winston, R. M., and Tuddenham, E. G. (1989) Biopsy of human preimplantation embryos and sexing by DNA amplification. *Lancet* **1,** 347–349.
19. Furrer, B., Candrian, U., Wieland, P., and Luthy, J. (1990) Improving PCR efficiency. *Nature* **346,** 324.
20. Longo, M. C., Berninger, M. S., and Hartley, J. L. (1990) Use of uracil DNA glycosylase to control carry-over contamination in polymerase chain reactions. *Gene* **93,** 125–128.

21. Pierce, K. E. and Wangh, L. J. (2004) Effectiveness and limitations of uracil-DNA glycosylases in sensitive real-time PCR assays. *Biotechniques* **36,** 44–46, 48.
22. Pang, J., Modlin, J., and Yolken, R. (1992) Use of modified nucleotides and uracil-DNA glycosylase (UNG) for the control of contamination in the PCR-based amplification of RNA. *Mol. Cell. Probes* **6,** 251–256.
23. Rys, P. N. and Persing, D. H. (1993) Preventing false positives: quantitative evaluation of three protocols for inactivation of polymerase chain reaction amplification products. *J. Clin. Microbiol.* **31,** 2356–2360.
24. Taggart, E. W., Carroll, K. C., Byington, C. L., Crist, G. A., and Hillyard, D. R. (2002) Use of heat labile UNG in an RT-PCR assay for enterovirus detection. *J. Virol. Methods* **105,** 57–65.
25. Kleiboeker, S. B. (2005) Quantitative assessment of the effect of uracil-DNA glycosylase on amplicon DNA degradation and RNA amplification in reverse transcription-PCR. *Virol. J.* **2,** 29.
26. Cimino, G. D., Metchette, K. C., Tessman, J. W., Hearst, J. E., and Isaacs, S. T. (1991) Post-PCR sterilization: a method to control carryover contamination for the polymerase chain reaction. *Nucleic Acids Res.* **19,** 99–107.
27. Fahle, G. A., Gill, V. J., and Fischer, S. H. (1999) Optimal activation of isopsoralen to prevent amplicon carryover. *J. Clin. Microbiol.* **37,** 261–262.
28. Zhu, Y. S., Isaacs, S. T., Cimino, C. D., and Hearst, J. E. (1991) The use of exonuclease III for polymerase chain reaction sterilization. *Nucleic Acids Res.* **19,** 2511.
29. Muralidhar, B. and Steinman, C. R. (1992) Geometric differences allow differential enzymatic inactivation of PCR product and genomic targets. *Gene* **117,** 107–112.
30. La, V. D., Clavel, B., Lepetz, S., Aboudharam, G., Raoult, D., and Drancourt, M. (2004) Molecular detection of Bartonella henselae DNA in the dental pulp of 800-year-old French cats. *Clin. Infect. Dis.* **39,** 1391–1394.
31. Lo, Y. M. D., Lo, E. S., Patel, P., Tse, C. H., and Fleming, K. A. (1991) Heteroduplex formation as a means to exclude contamination in virus detection using PCR. *Nucleic Acids Res.* **19,** 6653.
32. Yap, E. P., Lo, Y. M. D., Cooper, K., Fleming, K. A., and McGee, J. O. (1992) Exclusion of false-positive PCR viral diagnosis by single-strand conformation polymorphism. *Lancet* **340,** 736.
33. Fafi-Kremer, S., Brengel-Pesce, K., Bargues, G., et al. (2004) Assessment of automated DNA extraction coupled with real-time PCR for measuring Epstein-Barr virus load in whole blood, peripheral mononuclear cells and plasma. *J. Clin. Virol.* **30,** 157–164.
34. Knepp, J. H., Geahr, M. A., Forman, M. S., and Valsamakis, A. (2003) Comparison of automated and manual nucleic acid extraction methods for detection of enterovirus RNA. *J. Clin. Microbiol.* **41,** 3532–3536.
35. Cook, L., Ng, K. W., Bagabag, A., Corey, L., and Jerome, K. R. (2004) Use of the MagNA pure LC automated nucleic acid extraction system followed by real-time reverse transcription-PCR for ultrasensitive quantitation of hepatitis C virus RNA. *J. Clin. Microbiol.* **42,** 4130–4136.

3

Real-Time Polymerase Chain Reaction and Melting Curve Analysis

Robert J. Pryor and Carl T. Wittwer

Summary

Monitoring polymerase chain reaction (PCR) once each cycle is a powerful method to detect and quantify the presence of nucleic acid sequences and has become known as "real-time" PCR. Absolute quantification of initial template copy number can be obtained, although quantification relative to a control sample or second sequence is often adequate. Melting analysis following PCR monitors duplex hybridization as the temperature is changed and is a simple method for sequence verification and genotyping. Melting analysis is often conveniently performed immediately after PCR in the same reaction tube. The fluorescence of either DNA dyes that are specific to double-strands or fluorescently labeled oligonucleotide probes can be monitored for both real-time quantification and melting analysis. When used together with rapid temperature control, these methods allow amplification and genotyping in less than a half hour.

Key Words: Polymerase chain reaction (PCR); real-time PCR; melting curve analysis; fluorescent genotyping; nucleic acid quantification.

1. Introduction

Conventional polymerase chain reaction (PCR) requires that the product be analyzed after the reaction has finished *(1)*, a process which is often referred to as "end point" analysis. End point analysis usually requires a separation technique. For example, gels may be used to separate PCR products that are then stained with ethidium bromide. Although quantification is possible with end point analysis, it is not a simple task. Real-time instrumentation allows PCR quantification and analysis during amplification. In addition, end point melting analysis is often performed in the same tube as the PCR, providing a "closed tube" system for analysis (**Fig. 1**).

From: *Methods in Molecular Biology, vol. 336: Clinical Applications of PCR*
Edited by: Y. M. D. Lo, R. W. K. Chiu, and K. C. A. Chan © Humana Press Inc., Totowa, NJ

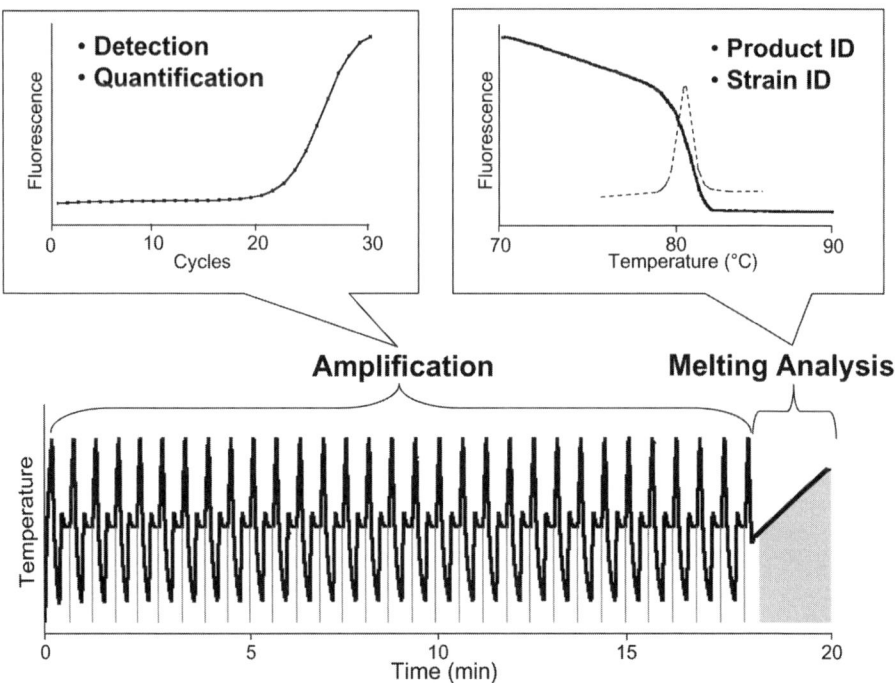

Fig. 1. Real-time monitoring during amplification and melting analysis. The bottom panel shows a typical rapid-cycle temperature profile that is followed by a temperature ramp for melting analysis. When the fluorescent signal is monitored during amplification once each cycle (dotted lines), it provides information on the presence or absence of specific target sequences and allows quantification of the target. When the fluorescent signal is monitored continuously through the melting phase (shaded area), it can provide information that verifies target identification, or establishes genotype. (Adapted from **ref. 2**, with the permission of ASM Press.)

The dynamic nature of real-time PCR and melting analysis simplifies detection, quantification, and genotyping. Fluorescent molecules are monitored with an optical thermocycler that provides fluorescent excitation and quantification of the fluorescent emission. The fluorophores may be covalently linked to an oligonucleotide to form a labeled primer or probe, or may be free molecules that bind to double stranded DNA. Many different designs are possible, the common feature being that they must exhibit a change in fluorescence during PCR so that product accumulation can be monitored.

The concentration of initial template in an optimized real-time PCR reaction determines how many cycles are necessary before the fluorescence rises. There is a direct analogy between the in vitro amplification of DNA by PCR and the in vivo proliferation of bacteria in a culture flask *(2)*. Initially, there is exponential amplification that is not observable because the concentrations are too low to be detected. This is followed by an observable growth phase, and finally a plateau phase in which growth is limited by nutrient consumption or overcrowding. The second derivative maximum (point of maximum acceleration) of this growth curve correlates with the initial template concentration. Specifically, this fractional cycle number is inversely proportional to the log of the initial template concentration. Such assays commonly have large dynamic ranges (5–8 decades), precision of 5–10%, excellent sensitivity (only limited by stochastic variation), and a specificity determined by the detection method and PCR quality.

If required, standards can be included to provide an exact copy number for absolute quantification. However, quantification relative to an experimental condition (mRNA) or natural reference (diploid DNA) usually provides the necessary information, eliminating the need for absolute standards. Sometimes a biological reference, such as a "housekeeping gene," is used to normalize results between experiments.

Melting curve analysis was originally used in conjunction with real-time PCR as presumptive identification of the target amplified *(3)*. SYBR Green I dye is included in the PCR reaction at concentrations that do not inhibit amplification. After PCR, the annealed products are melted at a constant rate (usually 0.1–0.3°C/s) and the decrease in fluorescence is monitored as the strands dissociate. Further work introduced probe melting analysis as a convenient genotyping method, first with two labeled oligonucleotides *(4,5)*, then with one *(6)*, and finally, with the development of saturating DNA dyes, unlabeled oligonucleotides *(7)*. High-resolution melting techniques expand the power of these techniques and can be performed in only 1–2 min *(8–10)*. When combined with rapid-cycle PCR *(11–15)*, amplification and analysis can easily be performed in less than 30 min.

2. Materials

2.1. Generic Reagents

For convenience and to decrease variation between reactions, it is good practice to premix all reagents that are in common; such a mixture is often referred to as a "master mix." If DNA (primers, probes, and template) is not included, these generic mixtures are stable for days at room temperature, weeks at 4°C, and months at –20°C. If a dye is included, the mixture should be kept in the dark. Although real-time PCR is robust and many formulations are possible, we recommend the following generic mixture. This mixture was developed for rapid-cycle PCR in glass capillaries with the LightCycler. If other sample containers are used, the bovine serum albumin (BSA) is not necessary.

5X Generic reagents	Concentration in final reaction
250 mM Tris, pH 8.3	50 mM
2.5 mg/mL BSA (*see* **Note 1**)	500 µg/mL
1 mM of each dNTP (dATP, dCTP, dGTP, and dTTP)	200 µM each
15 mM MgCl$_2$ (*see* **Note 2**)	3 mM
2 U/10 µL heat-stable polymerase (*see* **Note 3**)	0.4 U/10 µL
5X dye (*see* **Note 4**)	1X

2.2. Oligonucleotide Mixture

This mixture includes all primers and probes. For a 5X mixture, the primers are at 2.5 µM and probes at 1.0 µM, resulting in primer and probe concentrations of 0.5 µM and 0.2 µM, respectively, in the reaction (*see* **Note 5**). The oligonucleotide mixture is prepared in 1X TE (10 mM Tris, 0.1 mM ethylene-diamine tetraacetic acid [EDTA], pH 8.0) and is stable for weeks at 4°C and months at –20°C.

2.3. Template Solution

The typical amount of template per 10-µL reaction is approx 1.6×10^4 copies. This is approx 50 ng mammalian DNA (*see* **Note 6**), 50 pg bacterial DNA, 0.17 pg viral DNA, or 0.01 pg plasmid DNA. Template (in 1X TE') is usually diluted 2- to 10-fold in the final PCR solution. A fivefold dilution is assumed in the following. If absolute standards are used, it is best to dilute them from a concentrated stock solution (10^5–10^{10} copies per µL) just before use. Some templates (particularly diluted PCR products) are not stable for more than 24 h at 4–25°C. If RNA templates are of interest, reverse transcription is necessary (*see* **Note 7**).

3. Methods

3.1. Primer Design

Primer design for real-time PCR is similar to primer design for regular PCR. The following guidelines are useful:

1. Use primers between 15 and 30 bases that are matched in melting temperature (T_m) to each other. A primer greater than 17 bases long has a good chance of being unique in the human genome.
2. Unless there is a reason to amplify longer targets, choose a product length less than 500 bp, preferably less than 200 bp. Shorter products amplify with higher efficiency.
3. Avoid primers that anneal to themselves or to other primers, particularly at their 3' ends.

4. Choose primers that are specific to the target. Targets to avoid include pseudogenes (for genomic DNA) and related bacterial or viral strains (for microorganisms). Avoid simple sequence repeats and common repeated sequences, such as *Alu* repeats.
5. Avoid primers that have sequence complementary to internal sequences of the intended product.

As a final test, do a Basic Local Alignment Search Tool (BLAST) search (http://www.ncbi.nlm.nihgov/BLAST/) of the DNA that is likely to be present in your assay for sequences similar to those of your primers.

3.2. Probe Design

The easiest probe to design (and the least expensive) is one that is not needed. Before working with probes, try a SYBR Green I reaction. In most applications, the specificity of PCR complemented by melting analysis is adequate. If single-base genotyping is required, try small amplicon melting *(9)* or unlabeled probes *(7)*. Usually, covalently labeled fluorescent probes are only necessary with a bad PCR or when small template numbers must be quantified. Before moving to labeled probes, try one of the commercially available Hot Start techniques to improve specificity (*see* **Note 8**). Consider whether single-labeled probes will suffice *(4–6)* in lieu of more complex probes with two or more functional groups. The design of probes is covered extensively in the commercial literature and on many web sites.

3.3. Reaction Preparation (10-µL Reactions)

1. Prepare a final master mixture in a microfuge tube on ice by adding one volume of the 5X generic mixture and one volume the 5X oligonucleotide solution to two volumes of water. Mix the solution.

	For one 10-µL reaction	For *n* 10-µL reactions
Water	4 µL	4*n* µL
Generic reagents	2 µL	2*n* µL
Oligonucleotide solution	2 µL	2*n* µL

2. Add 8 µL of master mixture to each capillary.
3. Add 2 µL of each 5X template solution to each capillary. At this point, mixing is not necessary because it occurs during centrifugation and thermal cycling.
4. Transfer the capillaries to the LightCycler carousel and centrifuge briefly (alternatively, the capillaries may be centrifuged in individual adaptors and then transferred to the carousel).

3.4. Temperature Cycling for Real-Time PCR and Melting Analysis

Place the sample carousel into the LightCycler and program the desired temperature conditions for PCR (*see* **Note 9**). Begin thermal cycling and acquire fluorescence each cycle at the end of the extension phase. If melting will be performed in the LightCycler, the melting conditions are also entered before amplification. If a high-resolution melting instrument is available (*see* **Note 10**), the capillaries are transferred after PCR from the LightCycler to the high resolution instrument for melting.

3.5. Data Analysis

For absolute quantification, standards that bracket the potential concentration range are run in parallel. Typically, four to seven standards are separated in concentration by factors of 10. Replicates are not necessary, unless the instrument or preparation method are compromised or very low copy numbers (<20 copies per reaction) are being analyzed. A negative control (no template) should also be included in order to identify any background signal, although it is not used in quantitative analysis. Amplification plots or "growth curves" are displayed by plotting fluorescence vs cycle number (**Fig. 2**). A quantitative standard curve is constructed by relating the log of the initial number of templates in each standard to a fractional cycle number derived from each curve. Although a threshold fluorescence level can be used to define these fractional cycle numbers, the second derivative maximum is more precise and depends on the shape of the curve rather than on a specific fluorescence value *(2)*. Unknown concentrations are determined by interpolation from the standard curves.

For relative quantification, absolute standards are not necessary. The relative concentration of target between an experimental and a control sample depends on their fractional cycle numbers and the PCR efficiency (**Fig. 3**). For well optimized PCR reactions, the efficiency is close to 2.0 and the relative concentration depends only on the fractional cycle numbers. When greater accuracy is required, the PCR efficiency can be derived from the standard curve slope of template dilutions. In some experimental designs, the overall amount of nucleic acid (e.g., mRNA) may be difficult to control. In this case, it is common to normalize the test gene against a reference gene, such as a "housekeeping gene" that is assumed not to vary with the experimental conditions.

Melting analysis with SYBR Green I is commonly used to characterize PCR products *(3)*. A recent development is high-resolution melting analysis, enabling even single-base genotyping within PCR products more than 500 bp in length (**Fig. 4**). Samples are amplified in the presence of the dye LCGreen and melted at 0.3°C/s with 50–100 data points collected per degree Celsius.

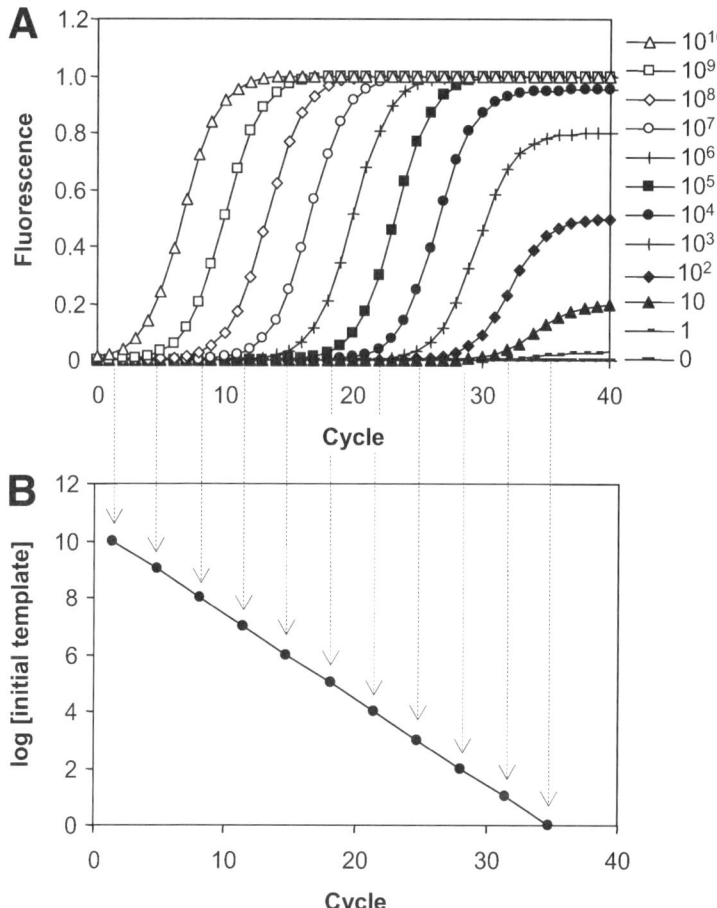

Fig. 2. Quantification by real-time polymerase chain reaction. Shown are typical real-time curves for amplification reactions of varying initial target concentrations (**A**), and the log of the initial concentration plotted against the cycle number at which the signal rises above background (**B**) as calculated by the second derivative maximum. (Adapted from **ref. 2**, with the permission of ASM Press.)

After fluorescence normalization, the shapes of the curves correlate with genotype. Melting domains are clearly identified as multiple melting transitions. This closed tube technique can be used for genotyping known mutations (*7–9*) and for scanning of unknown mutations (*16*).

Genotyping by melting curve analysis can be localized to a specific region and made more specific by limiting the amplicon length (*9*), or by using probes of various designs (**Fig. 5**). With short amplicons, the melting rate is kept at

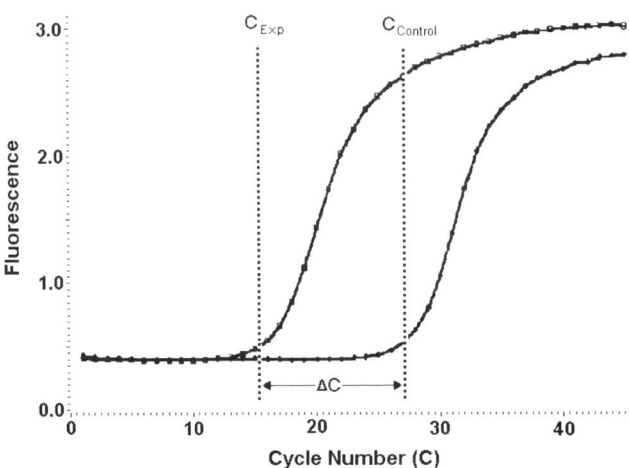

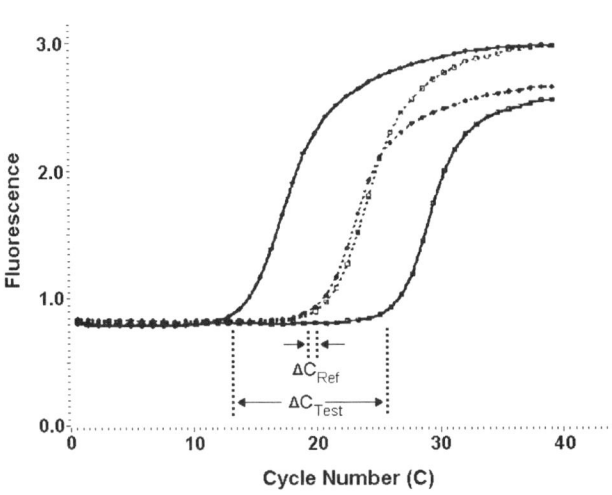

Fig. 3. Relative quantification by real-time polymerase chain reaction (PCR). (**A**) The amount of target in an experimental sample and a control sample are compared after PCR amplification and fluorescence monitoring at each cycle. For example, genomic DNA may be analyzed to assess gene amplification or deletion. Expression

Fig. 3. (*continued from opposite page*) of mRNA may also be studied after reverse transcription. The sample with the greater amount of DNA (or cDNA) will show an earlier increase in fluorescence. The second derivative maxima of the curves (vertical dotted lines) are determined as fractional cycle numbers. The relative copy number between samples is the PCR efficiency (eff) raised to the difference between fractional cycle numbers (ΔC). The calculation assumes that the PCR efficiency is the same between samples. The PCR efficiency is usually between 1.7 and 2.0. As a first approximation, an efficiency of 2 is often assumed. This analysis assumes that the starting amount of material (DNA or cDNA) in each sample is the same. **(B)** Another option is to use a test target normalized to a reference target. The amount of starting material in each sample is normalized to a reference (Ref) or housekeeping gene. Both experimental and control samples are amplified for both the test and reference targets. Any difference in the amount of starting material is normalized by the results of the reference target amplification. This method assumes that the reference target is invariant between samples and that the PCR efficiency for each target does not vary between samples. As a first approximation, an efficiency of 2 is often assumed for both targets and has become known as the ΔΔC method.

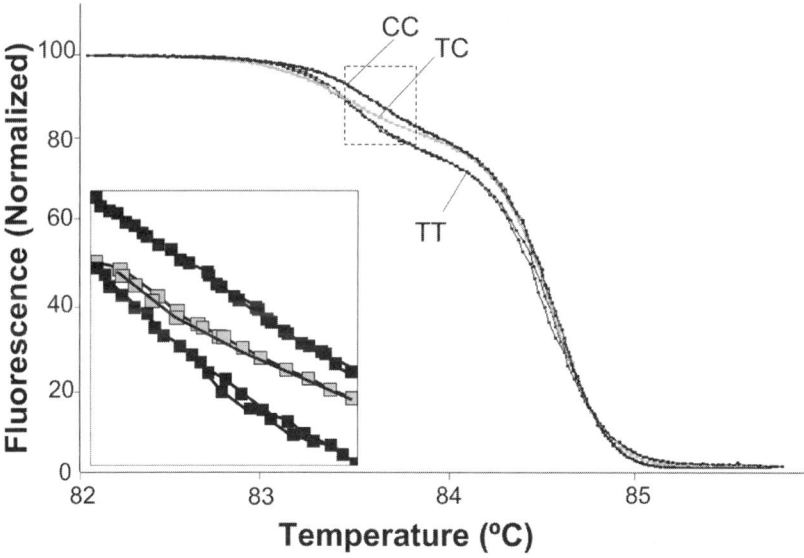

Fig. 4. A single-nucleotide polymorphism (SNP) is demonstrated in a 544-bp fragment by melting analysis. Shown are high-resolution melting curves of polymerase chain reaction amplicons from the *HTR2A* gene locus carrying an SNP. Results are shown for six individuals, two different individuals for each of the three genotypes: wild type homozygote (TT), mutant homozygote (CC), and heterozygote (TC). The inset is a magnified portion of the data showing that all three genotypes can be discriminated. (Adapted from **ref. 8**, with the permission of AACC Press.)

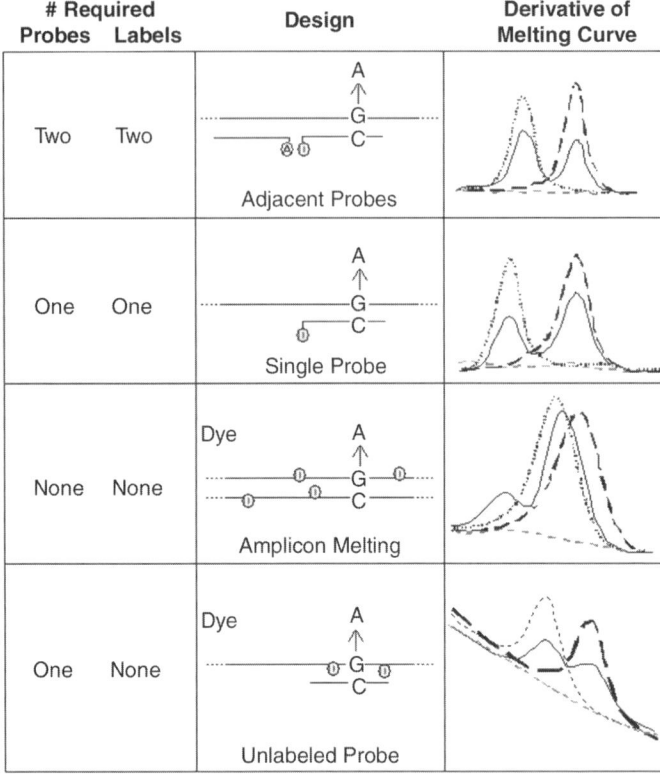

# Required Probes	Labels	Design	Derivative of Melting Curve
Two	Two	Adjacent Probes	
One	One	Single Probe	
None	None	Amplicon Melting	
One	None	Unlabeled Probe	

Fig. 5. Four modes of single nucleotide polymorphism genotyping by melting analysis. The traditional hybridization probe design (top row) uses a pair of probes, one labeled with an acceptor fluorophore (encircled A) and the other with a donor fluorophore (encircled D). The single-hybridization-probe design (second row) lacks the second probe. The amplicon melting design (third row) uses a saturating double-stranded DNA binding dye. The two homozygotes differ in melting temperature and the heterozygote has an additional low-temperature transition caused by heteroduplexes. The unlabeled probe design (bottom row), similar to amplicon melting, does not require a covalently attached fluorescent label and uses a DNA binding dye. However, because a probe is used, the derivative melting curves are better separated than with amplicon melting. Homozygous G allele (dashed line in far right column), homozygous A allele (dotted line), and the GA heterozygote (solid line). (Adapted from **ref. *17*** with the permission of Elsevier Press.)

0.3°C/s and high-resolution melting techniques may be necessary. When probes are used, the melting rate is usually 0.1°C and high resolution is not needed. Melting analysis using unlabeled probes and the dye LCGreen I is especially attractive because no covalently labeled oligonucleotides are required *(7)*.

4. Notes

1. BSA is used with PCR in glass capillary tubes to prevent polymerase denaturation on the glass surface. The grade of albumin is not critical, as long as it does not contain any DNA of interest and it is not acetylated. Siliconized tubes can also be used but are less convenient.

2. Most reactions work well with 3 mM MgCl$_2$, although a concentration range of 1–5 mM should be tried, along with annealing temperatures of 50–65°C, depending on the T$_m$ of the primers. For very high-T$_m$ products, additives such as dimethylsulfoxide (DMSO), formamide, or glycerol may be necessary for adequate denaturation.

3. Any native or engineered heat-stable polymerase can be used, although their extension rates may differ. Heat-activated polymerases require time for activation and increase the time required for PCR. 5'-exonuclease activity is required for some probe systems that depend on probe hydrolysis. Some 3'-exonuclease activity is useful for amplification of products longer than 1 kb for removal of incorporation errors.

4. The first dye used in real-time PCR was ethidium bromide *(18)*, an intercalating dye that binds between the bases of dsDNA. SYBR Green I was introduced to real-time PCR in 1997 *(19)* and is the dye most widely used today. SYBR Green I fluorescence is greater than that of ethidium bromide *(2)* and it can be viewed in the same channel as fluorescein, a common probe label. SYBR Green I is available as a 10,000X solution in DMSO from the manufacturer (Molecular Probes). Positive displacement pipets are necessary in order to accurately pipet small volumes of the stock solution in DMSO. This measurement is critical, because whereas a 1X solution gives maximal signal, a 2X solution completely inhibits PCR. A new class of "saturating" dyes that are not as prone to PCR inhibition has recently been developed (LCGreen, Idaho Technology) and enables high-resolution genotyping and scanning techniques.

5. Many different probes can be used in real-time PCR. They can be divided into two classes based on their mechanisms of action: hybridization and hydrolysis. The fluorescence of hybridization probes depends on whether the probe is bound to its target sequence or not. Fluorescence is reversible and melting curves can be obtained. Hydrolysis probes irreversibly change in fluorescence when the 5'-exonuclease activity of the polymerase cleaves the probe, separating the fluorescent label from the quencher. The relative merits of different probes are vigorously debated and adequately covered in commercial literature.

6. A DNA solution of 50 ng/µL (50 µg/mL) has an absorbance of 1.0 at 260 nm. This is a convenient 10X concentration of human genomic DNA for PCR.

7. Reverse transcription may be performed in the same solution as PCR (one-step) or separately (two-step). One-step reactions may use a bifunctional enzyme that accepts both RNA and DNA as templates, or two specific enzymes. One-step reactions use PCR primers for reverse transcription and are common in clinical assays, in which simplicity is paramount. In the research setting, two-step reactions that reverse-transcribe all mRNA with poly dT or random hexamers and

allow quantification of many transcripts from one reverse transcription may be preferred *(20)*.

8. Inappropriate annealing and extension can occur before PCR begins. Hot Start techniques delay the activation of the polymerase until high temperatures are reached. There are three common types of Hot Start methods: physical separation, antibodies, and heat-labile structural inactivation of the polymerase.

9. If the template is genomic DNA and has not been previously denatured, an initial denaturation of 10 s at 95°C is more than adequate. The only exception is when heat-activated polymerases are used, in which case the instructions from the manufacturer should be followed.

   For each cycle, there is no advantage to denaturation times greater than "0" s; that is, the denaturation temperature does not need to be held. Again, the only possible exception is when heat-activated polymerases are used. The fact that most PCR protocols use denaturation times of 5–60 s only reflects the poor heat transfer of conventional instruments *(17)*. Melting analysis with SYBR Green I can be used to determine the product T_m and establish necessary denaturation temperatures.

   Optimal annealing temperatures depend on the T_m of the primers, but usually range from 50 to 60°C for 20-mers with 3 mM $MgCl_2$. With primer concentrations of 0.5 µM, annealing is rapid and "0"-s holds can often be used. Specificity improves as the annealing time is decreased *(21)*.

   Extension times and temperatures depend on the target amplified. If the target is AT-rich, extension temperatures of less than 70°C may be required, whereas extension of GC-rich targets is faster at higher temperatures (approaching 80°C). For most targets, extension temperatures of 70–74°C work well and are most commonly used. Extension times depend on the length of the target. For products less than 100 bp, extension times of "0" s are adequate, even with rapid-cycling instruments. Products less than 200 bp should require no more than 10 s of extension. An extension time of 15–20 s may be required for products up to 500 bp, whereas 30–60 s will more often amplify a 1000-bp segment.

   Transition rates are usually programmed at 20°C/s, although a slower rate (1–2°C/s) between annealing and extension can improve yield in some cases. In real-time PCR (as well as in conventional PCR) there is no reason for a long final extension.

   Before melting analysis, the products should be denatured (95°C for 0 s) and rapidly cooled (–20°C/s) to 10°C below the lowest expected melting transition. Melting analysis on the LightCycler can be performed immediately. Alternatively, the samples can be stored at 4–25°C for at least 24 h before analysis on separate (e.g., high-resolution) instruments.

10. High-resolution fluorescent melting instrumentation *(7–10)* can be obtained through Idaho Technology (HR-1 and LightScanner).

References

1. Saiki, R. K., Gelfand, D. H., Stoffel, S., et al.(1988) Primer-directed enzymatic amplification of DNA with a thermostable DNA polymerase. *Science* **239,** 487–491.

2. Wittwer, C. T. and Kusukawa, N. (2004) *Real-time PCR, in Diagnostic Molecular Microbiology: Principles and Applications* (Persing D. H., Tenover F. C., Relman D. A., et al., eds.). ASM, Washington DC: pp. 71–84.

3. Ririe, K. M., Rasmussen, R. P., and Wittwer, C. T. (1997) Product differentiation by analysis of DNA melting curves during the polymerase chain reaction. *Anal. Biochem.* **245,** 154–160.

4. Lay, M. J. and Wittwer, C. T. (1997) Real-time fluorescence genotyping of factor V Leiden during rapid-cycle PCR. *Clin. Chem.* **43,** 2262–2267.

5. Bernard, P. S., Ajioka, R. S., Kushner, J. P., and Wittwer, C. T. (1998) Homogeneous multiplex genotyping of hemochromatosis mutations with fluorescent hybridization probes. *Am. J. Pathol.* **153,** 1055–1061.

6. Crockett, A. O. and Wittwer, C. T. (2001) Fluorescein-labeled oligonucleotides for real-time PCR: using the inherent quenching of deoxyguanosine nucleotides. *Anal. Biochem.* **290,** 89–97.

7. Zhou, L., Myers, A. N., Vandersteen, J. G., Wang, L., and Wittwer, C. T. (2004) Closed-tube genotyping with unlabeled oligonucleotide probes and a saturating DNA dye. *Clin. Chem.* **50,** 1296–1298.

8. Wittwer, C. T., Reed, G. H., Gundry, C. N., Vandersteen, J. G., and Pryor, R. J. (2003) High-resolution genotyping by amplicon melting analysis using LCGreen. *Clin. Chem.* **49,** 853–860.

9. Liew, M. A., Pryor, R., Palais, R., et al. (2004) Genotyping of single nucleotide polymorphisms by high-resolution melting of small amplicons. *Clin. Chem.* **50,** 1156–1164.

10. Zhou, L., Vandersteen, J., Wang, L., et al. (2004) High resolution DNA melting curve analysis to establish HLA genotypic identity. *Tissue Antigens* **64,** 156–164.

11. Wittwer, C. T., Ririe, K. M., Andrew, R. V., David, D. A., Gundry, R. A., and Balis, U. J. (1997) The LightCycler: a microvolume multisample fluorimeter with rapid temperature control. *Biotechniques* **22,** 176–181.

12. Meuer, S. C., Wittwer, C., and Nakagawara, K. (2001) *Rapid cycle real-time PCR—methods and applications* . Springer, Berlin: p. 408.

13. Reischl, U., Wittwer, C., and Cockerill, F. (2002) *Rapid cycle real-time PCR- methods and applications : microbiology and food analysis* . Springer, Berlin: p. 258.

14. Dietmaier, W., Wittwer, C., and Sivasubramanian, N. (2002) *Rapid cycle real-time PCR - methods and applications : genetics and oncology* . Springer, Berlin: p. 205.

15. Wittwer, C.T., Hahn, M., and Kaul, K. (2004) *Rapid cycle real-time PCR—methods and applications: quantification* . Springer, Berlin: p. 223.

16. Reed, G. H. and Wittwer, C. T. (2004) Sensitivity and specificity of single-nucleotide polymorphism scanning by high-resolution melting analysis. *Clin. Chem.* **50,** 1748–1754.

17. Wittwer, C. T. and Kusukawa, N. (2005) Nucleic Acid Techniques, in *Tietz Textbook of Clinical Chemistry and Molecular Diagnostics* , 4th Ed. (Burtis C. A., Ashwood E. R., and Bruns D. E., eds.). Elsevier, Philadelphia: pp. 1407–1449.

18. Higuchi, R., Dollinger, G., Walsh, P. S., and Griffith, R. (1992) Simultaneous amplification and detection of specific DNA sequences. *Biotechnology (NY)* **10,** 413–417.

19. Wittwer, C. T., Herrmann, M. G., Moss, A. A., and Rasmussen, R. P. (1997) Continuous fluorescence monitoring of rapid cycle DNA amplification. *Biotechniques* **22,** 130–131, 134–138.

20. Morrison, T. B., Weis, J. J., and Wittwer, C. T. (1998) Quantification of low-copy transcripts by continuous SYBR Green I monitoring during amplification. *Biotechniques* **24,** 954–958, 960, 962.

21. Wittwer, C. T. and Garling, D. J. (1991) Rapid cycle DNA amplification: time and temperature optimization. *Biotechniques* **10,** 76–83.

4

Qualitative and Quantitative Polymerase Chain Reaction-Based Methods for DNA Methylation Analyses

Ivy H. N. Wong

Summary

DNA methylation can be analyzed easily by qualitative or quantitative polymerase chain reaction (PCR)-based methods, including methylation-specific PCR (MSP), bisulfite sequencing, methylation-sensitive restriction enzyme PCR, combined bisulfite restriction analysis (COBRA), methylation-sensitive single nucleotide primer extension (Ms-SNuPE), and quantitative real-time MSP. MSP, which couples the bisulfite modification of DNA and PCR, is fast, highly sensitive, specific, and widely applied for DNA methylation analyses. Bisulfite modification converts unmethylated cytosine to uracil, whereas methylcytosine remains unmodified. Most of these methods require specific PCR primers that are designed to distinguish between methylated and unmethylated DNA sequences. Bisulfite sequencing is comparatively time-consuming. Methylation-sensitive restriction enzyme PCR combines methylation-sensitive restriction enzyme digestion and PCR. After enzyme digestion, PCR products are obtained if the enzyme does not digest the methylated CpG sites within the specified DNA region. COBRA, Ms-SNuPE, and quantitative real-time MSP allow the quantitative analyses of DNA methylation.

Key Words: DNA methylation; methylation-specific PCR (MSP); bisulfite sequencing; methylation-sensitive restriction enzyme PCR; combined bisulfite restriction analysis (COBRA); methylation-sensitive single nucleotide primer extension (Ms-SNuPE); quantitative real-time MSP.

1. Introduction

DNA methylation takes place after DNA synthesis by the enzymatic transfer of a methyl group from the methyl donor S-adenosylmethionine to the carbon-5 position of cytosine. Cytosines (Cs) usually located 5' to guanosines (Gs) are differentially methylated in the human genome (*1*). Non-CpG-rich sequences are interspersed with CpG islands, which are approx 500 bp long with G to C contents 55% and observed over expected frequencies of CpG

From: *Methods in Molecular Biology, vol. 336: Clinical Applications of PCR*
Edited by: Y. M. D. Lo, R. W. K. Chiu, and K. C. A. Chan © Humana Press Inc., Totowa, NJ

dinucleotides 0.65. The CpG islands of an increasing number of human genes are known to be differentially methylated in human tissues.

DNA methylation can be easily analyzed by qualitative or quantitative polymerase chain reaction (PCR)-based methods *(2)*, including methylation-specific PCR (MSP) *(3)*, bisulfite sequencing *(4,5)*, methylation-sensitive restriction enzyme PCR, combined bisulfite restriction analysis (COBRA) *(6)*, methylation-sensitive single nucleotide primer extension (Ms-SNuPE) *(7)*, and quantitative real-time MSP *(8)*. **Table 1** enumerates the advantages and disadvantages of these various techniques for DNA methylation analyses.

MSP, which couples the bisulfite modification of DNA and PCR, is fast, highly sensitive, relatively simple, inexpensive, and widely applied for DNA methylation analyses. Bisulfite modification converts unmethylated cytosine to uracil, whereas methylcytosine remains unmodified *(4,5)*. MSP requires specific primer sets, which are designed to distinguish between methylated and unmethylated DNA sequences. MSP offers high sensitivity for detecting small amounts of methylated alleles in clinical samples, such as plasma, serum, blood cells, lymph nodes, biopsies, and paraffin-embedded tissues *(9–11)*.

Bisulfite sequencing is comparatively time-consuming. Large-scale sequencing of multiple plasmid clones is required in order to obtain the overall methylation pattern *(4,5)*. Methylation-sensitive restriction enzyme PCR combines methylation-sensitive restriction enzyme digestion and PCR *(12)*. After enzyme digestion, PCR products are obtained if the enzyme does not digest at the methylated CpG sites within the specified DNA region. COBRA, Ms-SNuPE, and quantitative real-time MSP allow the quantitative analyses of DNA methylation. However, these methods are rather costly or are not easily established.

2. Materials

2.1. MSP, Methylation-Sensitive Restriction Enzyme PCR, COBRA, MsSNuPE, and Quantitative Real-Time MSP

1. Qiagen DNA extraction kit (Qiagen, Hilden, Germany).
2. CpGenome DNA modification kit (Chemicon International Inc., Temecula, CA) including sodium bisulfite reagents and carrier DNA.
3. Sodium hydroxide.
4. β-mercaptoethanol.
5. TE (Tris + ethylenediamine tetraacetic acid [EDTA]) buffer.
6. Ethanol.
7. Reaction buffers for PCR, restriction digestion and primer extension.
8. Primers (*see* **Note 1**).

Table 1
Advantages and Disadvantages of Methods for DNA Methylation Analyses

Method	Advantages	Disadvantages
MSP	• Highly sensitive • Highly specific for particular CpG sites • Facilitates the analysis of clinical samples with low levels of methylated sequences • Obviates the use of restriction enzymes and eliminates the problem of incomplete enzyme digestion	• Gives rise to false-positivity if bisulfite modification is incomplete • Poor design of primers can give rise to inconclusive results
Bisulfite sequencing	• Highly specific • Delineates the methylation status of each individual CpG site	• Technically demanding • Time-consuming and labor intensive
Methylation-sensitive restriction enzyme PCR	• More sensitive than Southern blot analysis	• Gives rise to false-positivity if bisulfite modification is incomplete • Gives rise to false-positivity if enzyme digestion is incomplete • Detects methylated CpG sites within methylation-sensitive restriction sites
COBRA	• Quantitative analysis of the methylation status of individual CpG sites	• Incomplete enzyme digestion can give rise to false-positivity
Ms-SNuPE	• Quantitative analysis of the methylation status of single CpG sites	• Handling of radioisotopes
Quantitative real-time MSP	• Quantitative analyses of the methylation status of multiple CpG sites within a particular DNA region of interest • Gel electrophoresis is not required • Requires optimization	• Needs optimization • Poor design of primers and probes can give rise to inconclusive results

PCR, polymerase chain reaction; MSP, methylation-specific PCR; COBRA, combined bisulfite restriction analysis; Ms-SNuPE, methylation-sensitive single nucleotide primer extension.

2.2. Enzymes

1. GeneAmp DNA amplification kit including Ampli*Taq* Gold polymerase (Applied Biosystems, Foster City, CA).
2. Methylation-sensitive restriction enzymes such as *Sma*I, *Eag*I, and *Sac*II, which recognize specific CpG-rich sequences for digestion (New England Biolabs, Beverly, MA).

2.3. Radioisotopes for MsSNuPE

[^{32}P] dCTP, [^{32}P] dTTP, [^{32}P] dATP, and [^{32}P] dGTP.

2.4. Quantitative Real-Time MSP

Fluorescent probes (*see* **Note 1**).

2.5. Equipment and Apparatus

1. Heat block or water bath.
2. PCR thermocycler.
3. Vertical or horizontal gel electrophoresis apparatus.
4. Gel documentation system.
5. PhosphorImager.
6. Applied Biosystems 7700 Sequence Detector.

3. Methods (*see* Note 2)

3.1. Methylation-Specific PCR (MSP)

Based on DNA methylation abnormalities, MSP (*see* **Note 3**) is particularly useful for detecting circulating tumor DNA in the plasma/serum and circulating tumor cells in the blood of cancer patients *(9,13)*. Aberrant promoter methylation has increasingly become a fundamental molecular abnormality leading to transcriptional silencing of tumor suppressor genes, DNA repair genes, and metastasis inhibitor genes *(14)*. Of significance, DNA hypermethylation of multiple genes successfully detected in circulating tumor DNA or cells from cancer patients may prove valuable for cancer detection and monitoring *(15)*. A number of methylation markers may allow the detection of circulating tumor DNA or cells from patients with different cancer types *(16)*.

The methylated and unmethylated DNA sequences are different after bisulfite modification, by which cytosine residues are deaminated to uracil residues. MSP primers are designed so that different primer sets can specifically anneal to sequences that contain methylated cytosines or modified thymines in the CpG sites within the target DNA sequence *(3)*. Careful design of primers is very important for eliminating or reducing the possibility of false-positivity or false-negativity (*see* **Note 4**). Incomplete bisulfite modification of DNA can give rise to false-positivity for methylated cytosines, which are actually unmodified cytosines. The sensitivities of MSP for various genes are very different, ranging from 10^{-5} to 10^{-2} *(9,10,17,18)*. The lower detectability of methylation changes may possibly be related to the lower sensitivity of the MSP assay. The detectability of methylation changes can possibly be further enhanced if a larger amount of DNA is added for PCR. Highly specific and sensitive MSP should be applicable for the methylation analysis among the low percentage of target cells and particularly useful for detecting small amounts of methylated target alleles.

3.1.1. DNA Extraction

The quality of DNA extracted from different sample sources is the determining factor affecting the accuracy of DNA methylation analysis. Also, the DNA quantity should be considered when determining which PCR-based method should be used for the methylation analysis. In particular, when the DNA amount is less than 1 µg, carrier DNA should be added to the DNA solution before the bisulfite modification steps.

3.1.2. Bisulfite Modification of DNA (see **Note 5**)

1. Add 7.0 µL of 3 *M* sodium hydroxide to 1 µg DNA in 100 µL of autoclaved distilled water and incubate the mixture at 37°C for 10 min.
2. Add 550 µL of freshly prepared DNA Modification Reagent I (CpGenome DNA modification kit; Chemicon International Inc.) and vortex.
3. Incubate at 50°C for 16–20 h in a heat block or water bath.
4. Add 5 µL of well suspended DNA Modification Reagent III from the kit to the incubated reaction mix.
5. Add 750 µL of DNA Modification Reagent II and mix briefly.
6. Centrifuge at 5000*g* for 10 s.
7. Add 0.5 mL of 70% ethanol and vortex.
8. Centrifuge at 5000*g* for 10 s and discard the supernatant.
9. Repeat the wash with 70% ethanol for two more times.
10. Remove the supernatant and dry the sample for 5 min.
11. Resuspend the pellet with 50 µL of 20 m*M* sodium hydroxide/90% ethanol solution and incubate at room temperature for 5 min.
12. Centrifuge at 5000*g* for 10 s and remove the supernatant.
13. Add 1 mL of 90% ethanol and vortex.
14. Centrifuge at 5000*g* for 10 s and remove the supernatant.
15. Repeat **steps 10** and **11**.
16. Dry the DNA pellet and dissolve in 25 µL of TE buffer.
17. Incubate the sample at 50–60°C for 5 min to elute the DNA.
18. Centrifuge at 5000*g* for 10 s and collect the supernatant for storage at –20°C.

3.1.3. PCR Primers

As an example, the methylated and unmethylated primer sequences for analyzing *p15* methylation by MSP are listed in **Table 2 (10)**.

Table 2
Methylated and Unmethylated Primer Sequences for Analyzing *p15*
Methylation by Methylation-Specific Polymerase Chain Reaction

Methylated	p15MF	5' GCG TTC GTA TTT TGC GGT T 3'
Methylated	p15MR	5' CGT ACA ATA ACC GAA CGA CCG A 3'
Unmethylated	p15UF	5' TGT GAT GTG TTT GTA TTT TGT GGT T 3'
Unmethylated	p15UR	5' CCA TAC AAT AAC CAA ACA ACC AA 3'

3.1.4. PCR Conditions and Thermocycling Profiles

PCR amplification is carried out using reagents supplied in the GeneAmp DNA Amplification Kit and Ampli*Taq* Gold polymerase. The thermal profile consists of an initial denaturation step of 95°C for 12 min followed by repetitions of 95°C for 45 s, 60°C for 45 s, and 72°C for 1 min, with a final extension step of 72°C for 10 min *(9)*. A total of 35–45 cycles are generally used.

3.1.5. Molecular Analyses of PCR Products

PCR products are analyzed by gel electrophoresis using 6% polyacrylamide gel or 2% agarose gel stained with ethidium bromide. Each DNA sample should be analyzed at least in duplicate.

3.2. Bisulfite Sequencing

Bisulfite sequencing was first described by Frommer and others in 1992 *(5)*. Bisulfite-modified DNA is PCR-amplified, cloned, and then sequenced. Bisulfite sequencing of DNA containing CpG sites is an excellent method for analyzing the methylation status of each individual CpG site within the target sequence *(4)*. The method is based on the chemical modification of unmethylated cytosine to uracil by sodium bisulfite; methylated cytosine remains unmodified *(19)*. The modified DNA is then PCR-amplified using gene-specific primers that anneal to the DNA sequences without CpG sites. In other words, the primers anneal to the DNA sequences containing uracils in positions of cytosines after bisulfite conversion. The amplified DNA fragments can be sequenced after cloning of individual molecules *(20)*. It is very important to verify whether the bisulfite conversion is complete.

3.3. Methylation-Sensitive Restriction Enzyme PCR

Digestion of DNA with methylation-sensitive restriction enzymes is followed by PCR amplification using primers flanking the target region where the restriction sites are located. DNA is cleaved with methylation-sensitive or methylation-insensitive restriction enzymes. Using methylation-sensitive restriction enzymes, one would be able to amplify the DNA fragment of an expected size if the CpG dinucleotides within the recognition sequences are methylated and, hence, resistant to enzyme digestion *(12)*. Using methylation-insensitive restriction enzymes, no PCR products are obtained regardless of the methylation status of CpG sites within the restriction sites. The major limitation of this method is that the enzyme digestion must be complete *(2)* (*see* **Note 6**). Another limitation is that if several CpG-rich restriction sites are present in the target region (some CpG sites are methylated and some are unmethylated), the results obtained can be inconclusive (*see* **Note 7**).

3.4. Combined Bisulfite Restriction Analysis

After bisulfite modification and PCR, the conversion of unmethylated cytosines to thymines and the retention of methylated cytosines can generate new restriction sites or retain restriction sites such as "CGCG" for *Bst*UI digestion *(6)*. The PCR products are thus sensitive to digestion by methylation-sensitive restriction enzymes, which recognize and digest the methylated DNA sequences. For COBRA, PCR primer sequences do not contain CpG dinucle-otides, so PCR amplification does not discriminate between methylated and unmethylated DNA templates. Therefore, the resulting PCR products may consist of some DNA fragments with newly created or retained restriction sites containing CpG dinucleotides, indicating the relative amounts of DNA sequences with methylated CpG sites (digested PCR products) at particular restriction sites in the DNA fragment *(6)*. COBRA is thus a quantitative method for determining the methylation levels of particular CpG sites (*see* **Note 8**).

3.5. Methylation-Sensitive Single Nucleotide Primer Extension

Quantitative assessment of DNA methylation changes at specific CpG sites can be performed by Ms-SNuPE based on bisulfite modification of DNA and single nucleotide primer extension *(7)* (*see* **Note 8**). The single nucleotide primer extension assay was first described by Kuppuswamy and others for the detection of mutations in abnormal alleles *(21)*. Gonzalgo and Jones *(7)* modified this method for the quantification of DNA methylation differences at specific CpG sites. In contrast with COBRA, this method requires radioisotope incorporation.

DNA is first treated with sodium bisulfite to convert unmethylated cytosines to uracils, but methylcytosines remain as cytosines. After bisulfite treatment and PCR amplification of the target sequence with gene-specific primers, PCR products serve as templates for MS-SNuPE. The primers used for the single nucleotide extension are designed so that the primer ends immediately 5' of the single nucleotide (just before the incorporation site) designated for the methylation analysis.

Quantification of the ratio of methylated cytosines vs thymines (unmethy-lated cytosines) at the original CpG sites can be determined by incubating the gel-purified PCR products, primers, and *Taq* polymerase with either [^{32}P] dCTP or [^{32}P] dTTP followed by denaturing polyacrylamide gel electrophoresis and phosphorimage analysis. Opposite-strand Ms-SNuPE primers that would in-corporate either [^{32}P] dATP or [^{32}P] dGTP to assess the methylation status of individual CpG sites can also be designed. If the target CpG site is methylated, a C is incorporated during nucleotide extension. If the CpG site is unmethylated, a T is incorporated instead. Quantification of the incorporated C or T determines the methylation status of the particular CpG site. As discussed previously, primer design and complete modification of DNA are important in order to obtain consistent results.

3.6. Quantitative Real-Time MSP

Using a quantitative approach (*see* **Notes 3** and **8**), the biological implica-tions of DNA methylation changes can be clarified. Previously, real-time quan-titative MSP has been developed to demonstrate the biological significance of *p16* methylation index *(8,22)*. In this system, two amplification primers and a dual-labeled fluorogenic hybridization probe are included. One fluorescent dye serves as a reporter (fluorescein [FAM]), and its emission spectra are quenched by a second fluorescent dye (tetramethylrhodamine [TAMRA]). During the extension phase of PCR, the 5' to 3' exonuclease activity of the *Taq* DNA poly-merase cleaves the reporter from the probe, thus releasing it from the quencher, resulting in an increase in fluorescent emission at 518 nm. Three real-time MSP systems are required for the quantification of the bisulfite-converted methy-lated sequence, the bisulfite-converted unmethylated sequence, and the uncon-verted wild-type sequence of the target gene. DNA amplification can be carried out in a 96-well reaction plate format in an Applied Biosystems 7700 Sequence Detector. Calibration curves are set up in parallel with each analysis using ref-erence DNA samples with known methylation status.

The methylation index (%) in a sample is calculated using the following equation:

$$\text{Methylation index} = \frac{M}{M+U} \times 100\%$$

where *M* is the quantity of methylated sequences measured by real-time MSP after bisulfite conversion and *U* is the quantity of unmethylated sequences. The usefulness of this fluorescence-based real-time MSP has also been validated by two other groups who aimed at the quantification of *p16* and *hMLH1* methylation changes *(23,24)*.

4. Notes

1. The best design of primers and probes is crucial for virtually all qualitative or quantitative methods for DNA methylation analyses.
2. The methods described in this chapter are most widely applied for DNA methylation analyses. Other methods have also been developed for high-throughput DNA methylation analyses *(2)*; however, these are not described in this chapter for simplicity.
3. MSP is one of the most common methods for detecting DNA methylation because of its high sensitivity and specificity. Another advantage of MSP is that it enables qualitative analysis and can be easily upgraded for quantitative analysis using real-time PCR.
4. The potential problem of obtaining false-negativity in different sample sources can be avoided by using MSP and choosing various DNA regions for the DNA methylation analysis.
5. Bisulfite conversion of all unmethylated cytosines to uracils in DNA has been incorporated into many techniques for DNA methylation analyses *(12)*. These methods include MSP, quantitative real-time MSP, COBRA, MsSNuPE, and bisulfite sequencing. However, bisulfite modification of DNA must be complete; otherwise, false-positive results may be obtained.
6. MSP offers high specificity and has the advantage over methylation-sensitive restriction enzyme PCR, which can give rise to false-positivity as a result of incomplete restriction enzyme digestion.
7. It has been difficult to measure accurately the extent of DNA methylation at specified CpG sites within target sequences. Previously, methylation analysis has been based on methylation-sensitive restriction enzyme digestion, which allows the analysis of CpG dinucleotides limited within the restriction sites of a particular region. Also, this method does not allow an accurate analysis of DNA methylation in formalin-fixed material in which the DNA quality is sometimes not good enough.
8. Quantitative real-time MSP, COBRA, and MsSNuPE allow the quantitative assessment of DNA methylation changes at multiple or individual CpG dinucleotides.

References

1. Herman, J. G. and Baylin, S. B. (2003) Gene silencing in cancer in association with promoter hypermethylation. *N. Engl. J. Med.* **349,** 2042–2054.
2. Ushijima, T. (2005) Detection and interpretation of altered methylation patterns in cancer cells. *Nat. Rev. Cancer* **5,** 223–231.
3. Herman, J. G., Graff, J. R., Myohanen, S., Nelkin, B. D., and Baylin, S. B. (1996) Methylation-specific PCR: a novel PCR assay for methylation status of CpG islands. *Proc. Natl. Acad. Sci. USA* **93,** 9821–9826.
4. Dupont, J.M., Tost, J., Jammes, H., and Gut, I.G. (2004) De novo quantitative bisulfite sequencing using the pyrosequencing technology. *Anal. Biochem.* **333,** 119–127.

5. Frommer, M., McDonald, L. E., Millar, D. S., et al. (1992) A genomic sequencing protocol that yields a positive display of 5-methylcytosine residues in individual DNA strands. *Proc. Natl. Acad. Sci. USA* **89,** 1827–1831.

6. Xiong, Z. and Laird, P. W. (1997) COBRA: a sensitive and quantitative DNA methylation assay. *Nucleic Acids Res.* **25,** 2532–2534.

7. Gonzalgo, M. L. and Jones, P. A. (1997) Rapid quantitation of methylation differences at specific sites using methylation-sensitive single nucleotide primer extension (Ms-SNuPE). *Nucleic Acids Res.* **25,** 2529–2531.

8. Lo, Y. M. D., Wong, I. H. N., Zhang, J., Tein, M. S., Ng, M. H., and Hjelm, N. M. (1999) Quantitative analysis of aberrant p16 methylation using real-time quantitative methylation-specific polymerase chain reaction. *Cancer Res.* **59,** 3899–3903.

9. Wong, I. H. N., Lo, Y. M. D., Zhang, J., et al. (1999) Detection of aberrant p16 methylation in the plasma and serum of liver cancer patients. *Cancer Res.* **59,** 71–73.

10. Wong, I. H. N., Ng, M. H., Huang, D. P., and Lee, J. C. (2000) Aberrant p15 promoter methylation in adult and childhood acute leukemias of nearly all morphologic subtypes: potential prognostic implications. *Blood* **95,** 1942–1949.

11. Wong, I. H. N., Lo, Y. M. D., Yeo, W., Lau, W. Y., and Johnson, P. J. (2000) Frequent p15 promoter methylation in tumor and peripheral blood from hepatocellular carcinoma patients. *Clin. Cancer Res.* **6,** 3516–3521.

12. Yamada, Y., Watanabe, H., Miura, F., et al. (2004) A comprehensive analysis of allelic methylation status of CpG islands on human chromosome 21q. *Genome Res.* **14,** 247–266.

13. Wong, I. H. N., Chan, J., Wong, J., and Tam, P. K. (2004) Ubiquitous aberrant RASSF1A promoter methylation in childhood neoplasia. *Clin. Cancer Res.* **10,** 994–1002.

14. Esteller, M. (2005) Dormant hypermethylated tumour suppressor genes: questions and answers. *J. Pathol.* **205,** 172–180.

15. Baylin, S. and Bestor, T. H. (2002) Altered methylation patterns in cancer cell genomes: cause or consequence? *Cancer Cell* **1,** 299–305.

16. Laird, P. W. (2005) Cancer epigenetics. *Hum. Mol. Genet.* **14(Spec No 1),** R65–R76.

17. Esteller, M., Sanchez-Cespedes, M., Rosell, R., Sidransky, D., Baylin, S. B., and Herman, J. G. (1999) Detection of aberrant promoter hypermethylation of tumor suppressor genes in serum DNA from non-small cell lung cancer patients. *Cancer Res.* **59,** 67–70.

18. Grady, W. M., Rajput, A., Lutterbaugh, J. D., and Markowitz, S. D. (2001) Detection of aberrantly methylated hMLH1 promoter DNA in the serum of patients with microsatellite unstable colon cancer. *Cancer Res.* **61,** 900–902.

19. Hayatsu, H., Wataya, Y., and Kazushige, K. (1970) The addition of sodium bisulfite to uracil and to cytosine. *J. Am. Chem. Soc.* **92,** 724–726.

20. Clark, S. J., Harrison, J., Paul, C. L., and Frommer, M. (1994) High sensitivity mapping of methylated cytosines. *Nucleic Acids Res.* **22,** 2990–2997.

21. Kuppuswamy, M. N., Hoffmann, J. W., Kasper, C. K., Spitzer, S. G., Groce, S. L., and Bajaj, S. P. (1991) Single nucleotide primer extension to detect genetic diseases: experimental application to hemophilia B (factor IX) and cystic fibrosis genes. *Proc. Natl. Acad. Sci. USA* **88,** 1143–1147.
22. Wong, I. H. N., Zhang, J., Lai, P. B., Lau, W. Y., and Lo, Y. M. D. (2003) Quantitative analysis of tumor-derived methylated p16INK4a sequences in plasma, serum, and blood cells of hepatocellular carcinoma patients. *Clin. Cancer Res.* **9,** 1047–1052.
23. Eads, C. A., Danenberg, K. D., Kawakami, K., et al. (2000) MethyLight: a high-throughput assay to measure DNA methylation. *Nucleic Acids Res.* **28,** E32.
24. Jahr, S., Hentze, H., Englisch, S., et al. (2001) DNA fragments in the blood plasma of cancer patients: quantitations and evidence for their origin from apoptotic and necrotic cells. *Cancer Res.* **61,** 1659–1665.

5

In-Cell Polymerase Chain Reaction

Strategy and Diagnostic Applications

T. Vauvert Hviid

Summary

In situ polymerase chain reaction (PCR) refers to the amplification of specific nucleic acid sequences and subsequent visualization of the PCR products in tissue sections. When PCR is performed in fixed cells in suspension the term *in-cell PCR* is normally applied instead. For example, in-cell PCR has been used for intracellular detection of specific viruses. Furthermore, in mixtures of cells in suspension with different genotypes, in-cell PCR may be used for intracellular detection of DNA or mRNA sequences specific for one of the genotypes/population of cells. An in-cell PCR method is described that makes it possible to genotype a specific gene of interest derived from one individual in a mixture of cells from two individuals. This new method is based on in-cell PCR and dependent upon the presence of a well characterized and specific DNA sequence or polymorphism, which has to be present only in the cells from the individual being genotyped for the gene sequence of interest. The in-cell PCR method might find diagnostic applications in detection of intracellular pathogenic sequences in subgroups of cells and in noninvasive prenatal genetic diagnoses with the use of the few fetal cells circulating in maternal blood.

Key Words: In-cell PCR; *in situ* PCR; genotyping; cell mixtures; fetal cells; maternal blood; prenatal diagnosis.

1. Introduction

In situ polymerase chain reaction (PCR) refers to the use of the PCR method for amplification of specific nucleic acid sequences and subsequent visualization of the PCR products in tissue sections. In this way, the recognition of the presence of specific mRNA or DNA sequences in different types of cells and tissues can be obtained through the combined examination of PCR/reverse-transcription (RT)-PCR product signal and histology. When PCR is performed in fixed cells in suspension the term *in-cell PCR* is normally applied instead.

From: *Methods in Molecular Biology, vol. 336: Clinical Applications of PCR*
Edited by: Y. M. D. Lo, R. W. K. Chiu, and K. C. A. Chan © Humana Press Inc., Totowa, NJ

Although the use of *in situ* PCR has been hampered by problems of diffusion of PCR products and thereby false-positive results, this seems not to be a significant problem with cells in suspension *(1,2)*. In-cell PCR has been used for intracellular detection of human immunodeficiency virus (HIV) mRNA and DNA *(3)*. In mixtures of cells in suspension with different genotypes, in-cell PCR may be used for intracellular detection of DNA or mRNA sequences specific for one of the genotypes/population of cells *(4,5)*. The method described here shows that it is possible to genotype a specific gene of interest derived from one individual in a mixture of cells from two individuals. This new method is based on in-cell PCR and is dependent on the presence of a well characterized and specific DNA sequence or polymorphism, which must be present only in the cells from the individual being genotyped for the gene sequence of interest. Fetal cells, DNA, and RNA have been detected in maternal blood and plasma *(6–12)*. The in-cell PCR method might find diagnostic applications in detection of intracellular pathogenic sequences in subgroups of cells and in noninvasive prenatal genetic diagnoses with the use of the few fetal cells circulating in maternal blood. A model system of mixtures of male and female cells will be used to illustrate the principle of in-cell PCR for genotyping a specific polymorphic sequence in the male cells. Mixtures of cells from two different individuals are fixed and permeabilized in suspension. After coamplification of a DNA sequence specific for one of the individuals and the DNA sequence to be genotyped, the two PCR products are linked together in the fixed cells positive for both DNA sequences by complementary primer tails and further amplification steps. In mixtures of male and female CD71-positive cells from umbilical cord blood attached to immunomagnetic particles, a Y-chromosome-specific sequence *(TSPY)* is linked to a polymorphic Human Leukocyte Antigen *(HLA)-DPB1* sequence only in the male cells, leading to the correct *HLA-DPB1* genotyping of the male by DNA sequencing of a nested, linked *TSPY-HLA-DPB1* PCR product.

2. Materials

1. Cells in suspension (in this model system: umbilical cord blood cells).
2. Genomic DNA.
3. Immunomagnetic beads coupled with antibodies (anti-CD71 Dynabeads [Dynal; Norway]).
4. 1X PBS/2% FCS: 1X phosphate-buffered saline (PBS) containing 20 mL/L fetal calf serum (FCS).
5. 1X PBS.
6. Magnetic particle concentrator.
7. IntraStain reagent set (Dako; Denmark).
8. Oligonucleotide primers.
9. *Taq* polymerase (Life Technologies).

10. PCR buffer and reagents:
    a. 20 m*M* Tris-HCl (pH 8.4) or 10 m*M* Tris-HCl (pH 8.8).
    b. 50 m*M* KCl.
    c. MgCl$_2$.
    d. Deoxynucleotide triphosphate (dNTP).
11. DNA sequencing reagents.
12. Thermal cycler.
13. Automated DNA sequencing equipment and agarose gel equipment.

3. Methods

3.1. Overview of the General Principle

The main steps in the procedure are outlined briefly here. A detailed description follows in next sections. **Figures 1** and **2** illustrate the general principle of the in-cell linker PCR method. In the model system, CD71$^+$ cells are isolated from mixtures of male and female umbilical cord blood with use of anti-CD71 monoclonal antibody-coated immunomagnetic particles. The different steps of the procedure, which include several changes of the buffer in which the cells are suspended, can then be performed easily by immobilization of the beads binding the cells by use of a magnet (*see* **Note 1**). After fixation and permeabilization, the cells are resuspended in PCR buffer with the primers BIO5TSPY, A3TSPYHL, NY3DPB1, and A5DPB1HL. During the first PCR, two PCR products will be amplified in male cells: a sequence of the gene *TSPY* and a sequence of exon 2 of *HLA-DPB1* (**Fig. 1**). *TSPY (CYS14)* is a Y-chromosome-specific, single-copy gene encoding a testis-specific protein *(13)*, and was chosen as the male ("fetal")-specific marker sequence (*see* **Note 2**). The *HLA-DPB1* sequence was chosen as an example of a polymorphic gene of interest because it is a well characterized gene and approx 80 alleles have been reported so far (e.g., HLA Informatics Group, http://www.anthonynolan.org.uk /HIG; *[14]*); therefore, the probability that any two samples have some polymorphic differences in this sequence is large. Thus, *HLA-DPB1* sequences from either the male or female cells can easily be defined in most cases. At the end of the first PCR, the *TSPY* and the *HLA-DPB1* PCR products will begin to form a linked PCR product because of complementary primer tail sequences (**Figs. 1** and **2**). These tail sequences have no marked homology to any known human genome sequences. Differences in primer concentrations added will partly force this to happen. This first PCR and linkage of PCR products are supposed to occur to a great extent inside the fixed male cells. In the fixed female cells, only the *HLA-DPB1* sequence will be amplified from the genomic DNA. After the first PCR, the PCR mixture is removed from the cells, and the cells are washed once in 1X PCR buffer and resuspended in the second PCR mixture including the nested primers NTPY5 and N2DPB1 (**Fig. 1**). Removal of the

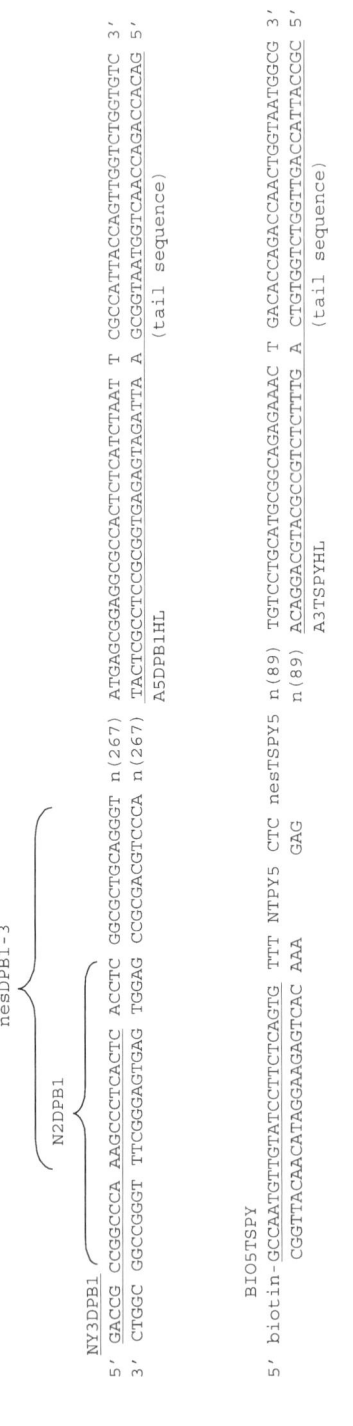

nesDPB1-3

N2DPB1

NY3DPB1
5' GACCG CCGGCCCA AAGCCCTCACTC ACCTC GGCGCTGCAGGGT n(267) ATGAGCGGAGGCGCCACTCTCATCTAAT T CGCCATTACCAGTTGGTCTGTGTC 3'
3' CTGGC GGCCGGGT TTCGGGAGTGAG TGGAG CCGCGACGTCCCA n(267) TACTCGCCTCCGCGGTGAGAGTAGATTA A GCGGTAATGGTCAACCAGACCACAG 5'
                                                                                      A5DPB1HL   (tail sequence)

BIO5TSPY
5' biotin-GCCAATGTTGTATCCTTCTCAGTG TTT NTPY5 CTC nesTSPY5 n(89) TGTCCTGCATGCGGCAGAGAAAC T GACACCAGACCAACTGGTAATGGCG 3'
          CGGTTACAACATAGGAAGAGTCAC AAA   GAG   n(89) ACAGGACGTACGCCGTCTCTTTG A CTGTGGTCTGGTTGACCATTACCGC 5'
                                                                      A3TSPYHL   (tail sequence)

5' biotin-GCCAATGTTGTATCCTTCTCAGTG n(150) TGTCCTGCATGCGGCAGAGAAAC T GACACCAGACCAACTGGTAATGGCG A 3' ◊
                                    ⇓ 3' A CTGTGGTCTGGTTGACCATTACCGC T TAATCTACTCTCACCG n(310) 5'
                                                                                    (tail sequence)

Fig. 1. Primer sequences and generation of a linked polymerase chain reaction (PCR) product. All primer sequences of the model system of the in-cell linker PCR principle are shown as parts of the two PCR products of *HLA-DPB1* and *TSPY*, respectively, amplified in the first PCR step. Below is the formation of the linked PCR product illustrated (n = number of nucleotides). An extra adenosine is included between the A5DPB1HL (or A3TSPYHL) primer sequence and the tail-linker sequence. This is because the *Taq* polymerase incorporates an adenosine on the 3' end of the new DNA strand.

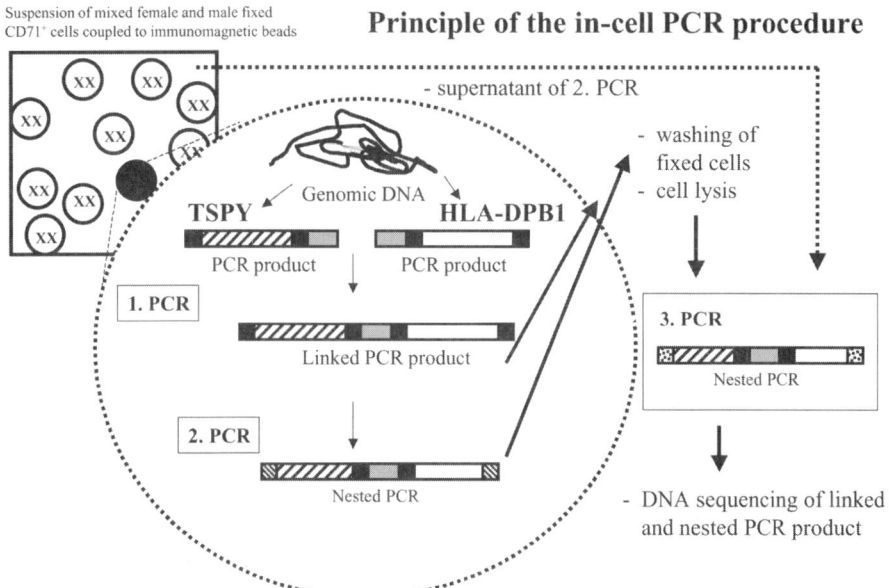

Fig. 2. Schematic drawing of the basic principle of the in-cell linker polymerase chain reaction.

first PCR mixture and washing of the fixed cells will remove any female-derived *HLA-DPB1* PCR products diffused out of fixed female cells that would be able to interfere with the following two nested PCR steps and the final DNA sequencing, by linkage to *TSPY* PCR products diffused out of male cells. During the second nested PCR, the linked *TSPY-DPB1* PCR product will be amplified further inside the fixed male cells (and outside cells in the supernatant). After the second PCR, the PCR mixture is removed (and also used as a template for a third PCR). Fixed cells are then resuspended and lysed in 1X PCR buffer; a portion of this "cell lysate" is then used as template for the third nested PCR using the primers nesTSPY5 and nesDPB1-3. This final PCR product can be visualized on a gel and directly DNA sequenced.

3.2. Initial Control Experiments With Extracted Genomic DNA

Before conducting the actual in-cell PCR linker PCR experiments, the performance of the co-amplification of the two PCR products and linkage has to be evaluated using extracted genomic DNA from males and females in a traditional PCR setting (primer sequences in **Fig. 1**). The method is based on the principle described by Diviacco et al. for constructing internal standards used in quantitative PCR techniques *(15)*.

3.2.1. Coamplification and Linkage of Two PCR Products

1. Two hundred nanograms of male or female genomic DNA is made up to a final volume of 50 μL containing 20 mM Tris-HCl (pH 8.4), 50 mM KCl, 2.5 mM MgCl$_2$, 200 μM deoxynucleotide triphosphate (dNTP), 50 pmol of each primer BIO5TSPY and NY3DPB1, 20 pmol of each primer A5DPB1HL and A3TSPYHL, and 2.5 U of *Taq* polymerase.
2. PCR conditions: denaturation for 5 min at 94°C, 35 cycles of annealing for 90 s at 55°C, extension for 90 s at 72°C, and denaturation for 45 s at 94°C.
3. One microliter of this first PCR product is used as template in a second PCR: Final volume of 50 μL with the same buffer conditions as the first PCR, except for 1.5 mM MgCl$_2$ and 25 pmol of primer NTPY5 and 25 pmol of primer N2DPB1. PCR conditions: 94°C for 2 min, 30 cycles of 94°C for 45 s, 55°C for 1 min, 72°C for 3 min, and a final extension of 72°C for 5 min.
4. Analyze amplified DNA fragments by electrophoresis on 2% agarose gels and stain with ethidium bromide.

3.2.2. Determination of a Lower Sensitivity Cutoff for the Co-Amplification Linker PCR Procedure

In a separate line of experiments to determine a lower sensitivity cutoff for the co-amplification linker PCR procedure using male genomic DNA, a third PCR is conducted.

1. Serial dilutions of male genomic DNA from 0.01 ng (corresponding to approx 1 cell) and to 5 ng (approx 500 cells) are used as templates.
2. The second PCR is modified in such a way that to the 50- μL volume of the first PCR, 50 μL of the second PCR mixture is added with a final amount of each of the primers NTPY5 and N2DPB1 of 100 pmol and 5 U of *Taq* polymerase.
3. Five microliters of this second PCR product is used as template in the third PCR: final volume of 50 μL containing 20 mM Tris-HCl (pH 8.4), 50 mM KCl, 1.5 mM MgCl$_2$, 200 μM dNTP, 40 pmol each of primers nesTSPY5 and nesDPB1-3, and 2.5 U of *Taq* polymerase. PCR conditions: denaturation for 2 min at 95°C; 40 cycles of denaturation for 45 s at 95°C, annealing for 60 s at 55°C, extension for 2 min at 72°C.
4. Analyze amplified DNA fragments by electrophoresis on 2% agarose gels and stain with ethidium bromide.

The two-step PCR procedure with co-amplification and linkage of the *TSPY* and *HLA-DPB1* PCR products followed by nested PCR in the traditional PCR setup produce clear bands after gel electrophoresis corresponding to the predicted length of the linked PCR product (531 bp) in only the male samples. Dilutions of male genomic DNA have shown that a concentration down to 0.05 ng, corresponding to genomic DNA from approximately seven cells, produces a positive PCR signal.

3.3. Isolation of CD71-Positive Umbilical Cord Blood Cells

The first steps in the actual in-cell PCR linker method described under **Subheading 3.3.1.** outline the procedure for specific isolation of a subgroup of cells in suspension.

3.3.1. Isolation of Specific Cell Populations With the Use of Immunomagnetic Beads

1. Process 1.0 or 0.5 mL of umbilical cord blood separately. Perform CD71$^+$ cell isolation from male or female samples alone or from mixtures (e.g., 1:1 or 1:9) of male and female samples made up to a volume of 1.0 or 5.0 mL. Perform the isolation of CD71$^+$ cells as described by the manufacture of the immunomagnetic beads.
2. Briefly, add 1.2×10^7 anti-CD71 Dynabeads (Dynal A/S, Norway) to 1 mL cord blood and incubate at 4°C with rotation for 20 min.
3. Add washing buffer (1X PBS/2% FCS) in a volume of 2–3 mL and place the tube in a magnetic particle concentrator (MPC) for 2–3 min.
4. Discard the supernatant and wash the captured cells three times in washing buffer and finally resuspend in 200 µL 1X PBS.

3.4. Fixation and Permeabilization of Cells in Suspension

Described as follows are the steps that can be used for fixation and permeabilization of the cells in suspension so the PCR reagents can enter the cells (*see* **Note 3**).

1. Place the CD71$^+$ cells bound to Dynabeads in 200 µL 1X PBS in the magnet and discard the supernatant.
2. Add 100 µL IntraStain Reagent A for fixation and resuspend the cells. Incubate for 15 min at room temperature.
3. Add to the cell suspension 1.4 mL 1X PBS and mix carefully. Discard the supernatant with the use of the magnet, and add 100 µL IntraStain Reagent B for permeabilization.
4. After 15 min of incubation at room temperature, add 1.4 mL 1X PBS and place the tube in the magnet and remove the supernatant.
5. Wash the fixed and permeabilized cells three times in 100 µL 1X PCR-buffer.

3.5. In-Cell Linker PCR of a Y-Chromosome-Specific DNA Sequence and an HLA-DPB1 DNA Sequence

The PCR steps in the in-cell linker PCR method are described under **Subheadings 3.5.1.–3.5.3.** First, the two PCR products of the Y-chromosome-specific DNA sequence and the polymorphic *HLA-DPB1* sequence are co-amplified and linked together in a subsequent PCR process. Thereafter, the two linked sequences are amplified further with the use of nested PCR.

3.5.1. PCR Co-Amplification and Linkage of the PCR Products

1. Resuspend fixed and permeabilized cells (approx 12–15,000 cells from 1 mL blood), derived from the CD71$^+$ cell isolation procedure, in 50 µL PCR mixture containing 20 m*M* Tris-HCl (pH 8.4), 50 m*M* KCl, 2.5 m*M* MgCl$_2$, 200 µ*M* dNTP, 100 pmol of each primer BIO5TSPY and NY3DPB1, 40 pmol of each primer A5DPB1HL and A3TSPYHL, and 5 U of *Taq* polymerase.
2. Perform PCR under the following conditions: denaturation for 5 min at 94°C, followed by 35 cycles of annealing 90 s at 55°C, extension for 90 s at 72°C, and denaturation for 45 s at 94°C.

3.5.2. The First Nested PCR

1. After the first PCR, place the tube in a magnet (MPC) and remove the supernatant. One hundred microliters of 1X PCR buffer was added for 1 min and removed, followed by the addition of 50 µL of the second PCR mixture: final volume of 50 µL with the same buffer conditions as the first PCR except for 2.5 m*M* MgCl$_2$ and 100 pmol each of primers NTPY5 and N2DPB1.
2. PCR conditions: 95°C for 2 min, followed by 35 cycles of 94°C for 45 s, 55°C for 90 s, 72°C for 90 s.

For some of the pure female samples, 2 µL of supernatant from the first PCR is used as template in a second PCR with 1.5 m*M* MgCl$_2$, 20 pmol each of N2DPB1 and 5NDPB1 (*see* **Table 1**), and 2.5 U of *Taq* polymerase; same PCR program. This is to obtain an *HLA-DPB1* exon 2 sequence of the female samples for later DNA sequencing. After the third PCR, the supernatant is removed using the magnet and stored at –20°C.

3.5.3. The Second Nested PCR

1. Resuspend the cells in 50 µL 1X PCR buffer and place them at 100°C for 8 min.
2. Two or ten microliters of the cell lysates or supernatants are used as template for the third nested PCR: total of 50 µL of PCR mixture containing 10 m*M* Tris-HCl (pH 8.8), 50 m*M* KCl, 1.5 m*M* MgCl$_2$, 0.8 µL/mL Nonidet P40, 200 µ*M* dNTP, 20 pmol each of primers nesTSPY5 and nesDPB1-3, and 2.5 U of *Taq* polymerase.
3. PCR conditions: denaturation for 2 min at 95°C, followed by 35 cycles of denaturation for 45 s at 95°C, annealing for 60 s at 55°C, and extension 2 min at 72°C. Amplified DNA fragments are analysed by electrophoresis on 2% agarose gels and stained with ethidium bromide.

Results from an in-cell linker PCR experiment performed on CD71$^+$ cells derived from a male and a female pregnancy and a 1:1 mixture of cord blood, respectively, are shown in **Fig. 3A**. Only pure male or mixed samples result in a linked and nested PCR product of the predicted size of 495 bp. Linked PCR products are present in the PCR supernatant after the third PCR step in samples

Table 1
List of Primers for Polymerase Chain Reaction Amplification[a]

BIO5TSPY	5' biotin-GCC AAT GTT GTA TCC TTC TCA GTG 3'
A3TSPYHL	5' <u>CGC CAT TAC CAG TTG GTC TGG TGT C A</u> G TTT CTC TGC CGC ATG CAG GAC A 3'
NY3DPB1	5' GAC CGC CGG CCC AAA GCC CTC ACT C 3'
A5DPB1HL	5' <u>GAC ACC AGA CCA ACT GGT AAT GGC G A</u> A TTA GAT GAG AGT GGC GCC TCC GCT CAT 3'
NTPY5	5' CTT CGG CCT TTC TAG TGG AGA GGT G 3'
N2DPB1	5' CCG GCC CAA AGC CCT CAC TCA CCT C 3'
nesTSPY5	5' TCG GGA AAG TGT AAG TAA CTG ATG GGC AGC 3'
nesDPB1-3	5' AAG CCC TCA CTC ACC TCG GCG CTG CAG GGT 3'
5NDPB1	5' GAG AGT GGC GCC TCC GCT CAT GTC CGC 3'
Y1NY5	5' CAG TGT GAA ACG GGA GAA AAC AGT 3'
Y2NY3	5' GTT GTC CAG TTG CAC TTC GCT GCA 3'
BY5NES	5' CAT GAA CGC ATT CAT CGT GTG GTC 3'
Y3NESNY	5' CTG CGG GAA GCA AAC TGC AAT TCT T 3'

[a]The linker tail sequence in primers A3TSPYHL and A5DPB1HL is underlined.

containing fixed male cells (**Fig. 3A**, lanes 5 and 6). There are no clear differences in the results of the third PCR based on supernatant or cell lysate, respectively, from the second PCR. The linked and nested PCR products cannot be visualized in ethidium bromide-stained gels after the second PCR. A third (and a second nested) PCR is necessary. This is also the experience of other in-cell PCR procedures *(16,17)*.

3.6. DNA Sequencing

DNA sequence PCR products with the use of automated DNA sequencing according to the manufacturer's instructions. Here is Dye Primer (Cy-5' labeling) technology used (a Thermo Sequenase fluorescence-labeled primer cycle sequencing method with 7-deaza-dGTP from Amersham Pharmacia Biotech, UK). Sequencing primers: DPB1SEK5, Cy-5' GCG CCT CCG CTC ATG TCC GCC CCC T 3', and TSPYSEK5, Cy-5' GTA AGT AAC TGA TGG GCA GCT CGG CT 3'.

Direct DNA sequencing results (electropherograms) of the PCR products are shown in **Fig. 3B–D**. Two *HLA-DPB1* polymorphisms, which were found to distinguish the male and female samples, are shown as examples. The *DPB1* exon 2 polymorphisms are well described (HLA Informatics Group; http://www.anthonynolan.org.uk/HIG) and are located respectively in codon 55 and codon 56. The female sample has GAT at codon 55 and GAG at codon 56 (**Fig. 3B**), whereas the male sample shows GCT at codon 55 and GCG at codon

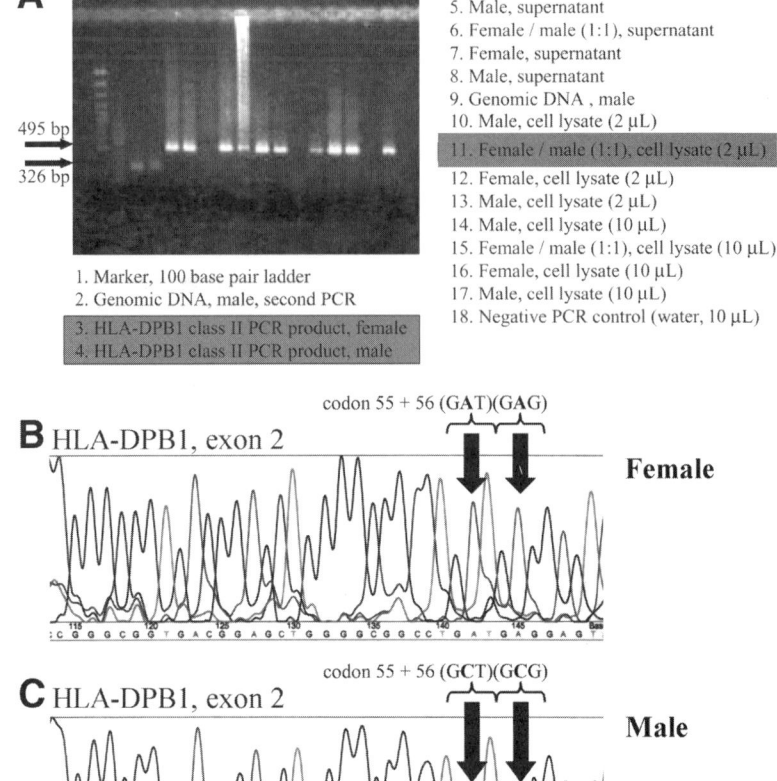

A

495 bp
326 bp

5. Male, supernatant
6. Female / male (1:1), supernatant
7. Female, supernatant
8. Male, supernatant
9. Genomic DNA , male
10. Male, cell lysate (2 µL)
11. Female / male (1:1), cell lysate (2 µL)
12. Female, cell lysate (2 µL)
13. Male, cell lysate (2 µL)
14. Male, cell lysate (10 µL)
15. Female / male (1:1), cell lysate (10 µL)
16. Female, cell lysate (10 µL)
17. Male, cell lysate (10 µL)
18. Negative PCR control (water, 10 µL)

1. Marker, 100 base pair ladder
2. Genomic DNA, male, second PCR
3. HLA-DPB1 class II PCR product, female
4. HLA-DPB1 class II PCR product, male

B HLA-DPB1, exon 2

codon 55 + 56 (GAT)(GAG)

Female

C HLA-DPB1, exon 2

codon 55 + 56 (GCT)(GCG)

Male

D HLA-DPB1, exon 2

codon 55 + 56 (GCT)(GCG)

**Mixture,
female and
male 1:1,
cell lysate**

Fig. 3. (**A**) Gel electrophoresis of the third linked and nested in-cell polymerase chain reaction (PCR) product from an experiment using a 1:1 mixture of umbilical cord blood from a male and a female pregnancy. (**B–D**) DNA sequencing results (electropherograms) of the PCR products shown in the two boxes in (**A**). The *HLA-DPB1* polymorphisms at codons 55 and 56 obtained from the lysate of the 1:1 mixture of male and female cord blood cells (**D**) are the ones of the male (**C**).

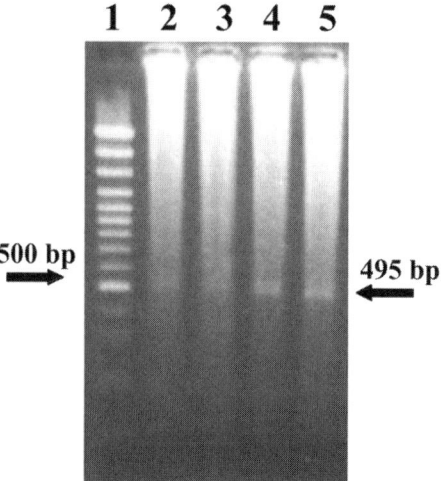

Fig. 4. Gel electrophoresis of the third linked and nested in-cell polymerase chain reaction product from an experiment involving the dilution of fixed and permeabilized CD71$^+$ cells derived from 0.5 mL of cord blood from a male pregnancy diluted in cells derived from 0.5 mL of cord blood from a female pregnancy in volume ratios of 1:10 (lane 2), 1:20 (lane 3), 1:50 (lane 4), and 1:250 (lane 5). Lane 1 is a DNA marker (100-bp ladder).

56 (**Fig. 3C**). The 1:1 mixture of male and female cells clearly shows the *HLA-DPB1* genotype of the male sample (**Fig. 3D**). This was also the case for other *HLA-DPB1* polymorphisms. Therefore, DNA sequencing of the PCR products clearly demonstrates that no cross-association has occurred between male and female PCR product sequences (*see* **Note 4**). So in mixtures of male and female cells, only the male *HLA-DPB1* polymorphic DNA sequence is detected by sequencing. In pure female samples, no linked PCR signal is obtained.

3.7. Studies of the Sensitivity of the In-Cell Linker PCR Method

To obtain data on the sensitivity of the in-cell linker PCR method fixed and permeabilized CD71$^+$ cells derived from 0.5 mL of cord blood from a male pregnancy can be diluted in the same type of cells derived from 0.5 mL of cord blood from a female pregnancy in volume ratios of 1:10, 1:20, 1:50, and 1:250. Afterward, the three PCR steps described previously are performed with the exception that the cells are lysed after the first PCR. Final PCR products are visualized on agarose gels. A positive in-cell linker PCR signal can be seen down to a ratio of 1:250 (*see* **Fig. 4**). Because 6-7000 CD71$^+$ cells were obtained from 0.5 mL of cord blood, a positive PCR signal may be obtained from at least approx 25 male cells in a male–female mixture of cells (*see* **Note 5**).

4. Notes

1. This was described earlier by Chapal et al. in another in-cell PCR procedure *(17)*. Microscopic examination of the enriched CD71$^+$ cell population coupled to immunomagnetic particles after fixation and permeabilization using the IntraStain reagents showed intact cells in conjunction with the beads. However, cells may be lost during the successive PCR steps. Centrifugation may be used for changes of buffer and reagents *(16)*; however, this did not work successfully in the current procedure. Experiments using centrifugation steps instead of Dynabeads for shifts of buffer or washing did not result in PCR signals; furthermore, there was a tendency in increasingly severe cell loss during these centrifugation steps. An alternative way to "immobilize" the fixed cells during the different steps in the procedure may be one way to improve the method.

2. Naturally, in one-half of all pregnancies, another fetal-specific marker DNA sequence must be used, such as the Rhesus D gene in cases of rhesus-negative mothers and rhesus-positive fetuses. In addition, many paternal/fetal specific DNA polymorphisms may be usable in an allele-specific PCR setup. Furthermore, recent studies have searched for and indicated the existence of more general specific fetal mRNA marker sequences which may be used in a RT-PCR-modified version of the in-cell PCR procedure presented here *(18,19)*.

3. Several different reagents have been used in the literature for fixation and permeabilization of cells in *in situ* or in-cell PCR protocols *(1,2)*. Initial experiments with 1% paraformaldehyde and proteinase K treatment did not result in any positive PCR results in the current method, but this does not mean that fixation and permeabilization reagents other than the IntraStain reagents will not work with the method.

4. The most pressing problem with the *in situ* or in-cell PCR technique, both in theory and apparently sometimes in practice, is the occurrence of false-positive results owing to diffusion of PCR products out of and/or into the fixed and permeabilized cells *(1)*. This would lead to a linkage of male (or fetal) marker sequences (in the model system the *TSPY* PCR product) and female (or maternal) PCR products of the gene sequence of interest (in the model system *HLA-DPB1* PCR products) in the described procedure. However, this seems not to be a problem in the presented model system; only male *HLA-DPB1* polymorphisms were detected in the mixture of male and female cells. In the literature, diffusion problems seem mainly to occur during *in situ* PCR in tissue sections, whereas the problem seems minimized with fixed cells in suspension *(1,2)*. One way proposed in the literature to reduce possible problems with diffusion is to use, e.g., biotin coupled to the primers or to include biotin- or digoxigenin-substituted nucleotides. The idea is that a more "bulky" PCR product accumulated inside the fixed cells will not diffuse out of the cells to the same extent as nonlabeled PCR products *(1)*. Another modification of the procedure could be to shorten the primer tail sequences in the first PCR and increase the length of the primers in the two following nested PCR steps. A high annealing temperature could thus be used in the second and third PCR, whereby no "false" *de novo* formation of linked PCR products would occur.

5. Interestingly, in the in-cell PCR study by Chapal et al. *(17)*, the best results of in-cell PCR amplification and linkage of the two PCR products were in the following order according to the number of cells included: 500 > 5000 > 50,000. This could indicate that the in-cell PCR efficiency is dependent on the number of fixed cells included in the reaction and may decrease with increasing cell number. There might be an optimum between the numbers of PCR-specific male cells, PCR-unspecific female cells, and magnetic beads. Interestingly, in dilution experiments a clearer band of linked PCR product was observed in the high dilution (1:250) of male cells in female cells compared with the low dilutions.

Acknowledgments

The author thanks Lone G. Nielsen for careful technical assistance on certain parts of this work.

References

1. Komminoth, P. and Long, A. A. (1993) In-situ polymerase chain reaction. An overview of methods, applications and limitations of a new molecular technique. *Virchows Arch. B. Cell Pathol. Incl. Mol. Pathol.* **64,** 67–73.
2. Long, A. A., Komminoth, P., Lee, E., and Wolfe, H. J. (1993) Comparison of indirect and direct in-situ polymerase chain reaction in cell preparations and tissue sections. Detection of viral DNA, gene rearrangements and chromosomal translocations. *Histochemistry* **99,** 151–162.
3. Patterson, B. K., Till, M., Otto, P., et al. (1993) Detection of HIV-1 DNA and messenger RNA in individual cells by PCR- driven in situ hybridization and flow cytometry. *Science* **260,** 976–979.
4. Timm, E. A. J., Podniesinski, E., Duckett, L., Cardott, J., and Stewart, C. C. (1995) Amplification and detection of a Y-chromosome DNA sequence by fluorescence in situ polymerase chain reaction and flow cytometry using cells in suspension. *Cytometry* **22,** 250–255.
5. Hviid, T. V. (2002) In-cell PCR method for specific genotyping of genomic DNA from one individual in a mixture of cells from two individuals: a model study with specific relevance to prenatal diagnosis based on fetal cells in maternal blood. *Clin. Chem.* **48,** 2115–2123.
6. Elias, S., Price, J., Dockter, M., et al. (1992) First trimester prenatal diagnosis of trisomy 21 in fetal cells from maternal blood. *Lancet* **340,** 1033.
7. Cheung, M. C., Goldberg, J. D., and Kan, Y. W. (1996) Prenatal diagnosis of sickle cell anaemia and thalassaemia by analysis of fetal cells in maternal blood. *Nat. Genet.* **14,** 264–268.
8. Bianchi, D. W., Simpson, J. L., Jackson, L. G., et al. (2002) Fetal gender and aneuploidy detection using etal cells in maternal blood: analysis of NIFTY I data. National Institute of Child Health and Development Fetal Cell Isolation Study. *Prenat. Diagn.* **22,** 609–615.
9. Lo, Y. M. D. (2005) Recent advances in fetal nucleic acids in maternal plasma. *J. Histochem. Cytochem.* **53,** 293–296.

10. Li, Y., Di Naro, E., Vitucci, A., Zimmermann, B., Holzgreve, W., and Hahn, S. (2005) Detection of paternally inherited fetal point mutations for beta-thalassemia using size-fractionated cell-free DNA in maternal plasma. *JAMA* **293,** 843–849.

11. Lo, Y. M. D., Corbetta, N., Chamberlain, P. F., et al. (1997) Presence of fetal DNA in maternal plasma and serum. *Lancet* **350,** 485–487.

12. Ng, E. K. O., Tsui, N. B. Y., Lau, T. K., et al. (2003) mRNA of placenta origin is readily detectable in maternal plasma. *Proc. Natl. Acad. Sci. USA* **100,** 4748–4753.

13. Arnemann, J., Jakubiczka, S., Thuring, S., and Schmidtke, J. (1991) Cloning and sequence analysis of a human Y-chromosome-derived, testicular cDNA, TSPY. *Genomics* **11,** 108–114.

14. Hviid, T. V., Madsen, H. O., and Morling, N. (1992) HLA-DPB1 typing with polymerase chain reaction and restriction fragment length polymorphism technique in Danes. *Tissue Antigens* **40,** 140–144.

15. Diviacco, S., Norio, P., Zentilin, L., et al. (1992) A novel procedure for quantitative polymerase chain reaction by coamplification of competitive templates. *Gene* **122,** 313–320.

16. Embleton, M. J., Gorochov, G., Jones, P. T., and Winter, G. (1992) In-cell PCR from mRNA: amplifying and linking the rearranged immunoglobulin heavy and light chain V-genes within single cells. *Nucleic Acids Res.* **20,** 3831–3837.

17. Chapal, N., Bouanani, M., Embleton, M. J., et al. (1997) In-cell assembly of scFv from human thyroid-infiltrating B cells. *Biotechniques* **23,** 518–524.

18. Hogh, A. M., Hviid, T. V., Christensen, B., et al. (2001) zeta-, epsilon-, and gamma-Globin mRNA in blood samples and CD71(+) cell fractions from fetuses and from pregnant and nonpregnant women, with special attention to identification of fetal erythroblasts. *Clin. Chem.* **47,** 645–653.

19. Boye, K., Hougaard, D. M., Ebbesen, P., Vuust, J., and Christiansen, M. (2001) Novel feto-specific mRNA species suitable for identification of fetal cells from the maternal circulation. *Prenat. Diagn.* **21,** 806–812.

6

Qualitative and Quantitative DNA and RNA Analysis by Matrix-Assisted Laser Desorption/Ionization Time-of-Flight Mass Spectrometry

Chunming Ding

Summary

Matrix-assisted laser desorption/ionization time-of-flight mass spectrometry (MALDI-TOF MS) gives extremely precise reading of mass-to-charge ratios (two analytes differ by 1 Da can be distinguished) and provides high sensitivity (less than 1 fmole of a DNA oligonucleotide can be detected), allowing DNA quantifications with single base specificity and single DNA molecule sensitivity (coupled with polymerase chain reaction [PCR]). To quantify a DNA sequence of interest, a competitive synthetic (60–80 bases) oligonucleotide standard with an artificial single base mutation in the middle is introduced, and these two virtually identical sequences are co-amplified by PCR. A third primer (extension primer) is designed to anneal to the region immediately upstream of the mutation site. Depending on the specific mutation introduced and the ddNTP/dNTP mixtures used, either one or two bases are added to the extension primer to produce two extension products from the two templates. Last, the two extension products are detected and quantified by high-throughput MALDI-TOF MS. In addition, with an improved primer extension method called single allele base extension reaction (SABER), rare mutant DNA can be robustly detected even when normal DNA is present at 50-fold or more than the DNA mutants.

Key Words: DNA mutation; single nucleotide polymorphism (SNP); gene expression; DNA quantification; matrix-assisted laser desorption/ionization time-of-flight mass spectrometry (MALDI-TOF MS); competitive PCR; primer extension; real-competitive PCR (rcPCR); single allele base extension reaction (SABER); prenatal diagnosis.

1. Introduction

Matrix-assisted laser desorption/ionization time-of-flight mass spectrometry (MALDI-TOF MS), originally developed for peptide analysis *(1)*, is extremely precise in determining the mass-to-charge ratios (single Dalton resolution *[2]*) and very sensitive (0.7 fmole DNA oligonucleotide *[3]*), yet remains poorly

From: *Methods in Molecular Biology, vol. 336: Clinical Applications of PCR*
Edited by: Y. M. D. Lo, R. W. K. Chiu, and K. C. A. Chan © Humana Press Inc., Totowa, NJ

quantitative. The poor quantitative ability of MALDI-TOF MS is mainly owing to ionization/desorption differences between amino acid residues (or nucleotide residues *[4]*), and uneven matrix-analyte crystal formations. MALDI-TOF MS is extremely fast (a few seconds for each sample) and has been highly automated *(3,5,6)*. As a result, MALDI-TOF MS is most widely used in high-throughput qualitative analyses, such as protein sequencing *(7)* and single nucleotide polymorphism (SNP) scoring *(3,8)*.

Although it is hard to quantify protein or DNA on the basis of absolute mass spectrometric signal intensity, it is highly accurate and reproducible to quantify the ratios of two chemically identical (or very similar) proteins or oligonucleotides in a single mass spectrum. Recently, MALDI-TOF MS has been used for high-throughput protein *(9)* and DNA/RNA *(10,11)* quantifications. This chapter provides a working protocol, which combines competitive polymerase chain reaction (PCR) and MALDI-TOF MS, for a DNA/RNA quantification technique known as real competitive PCR (rcPCR). In addition, DNA mutation analyses can be carried out simultaneously.

To quantify a DNA sequence of interest, a synthetic (60–80 bases) oligonucleotide standard with an artificial single base mutation in the middle is co-amplified with a virtually identical sequence of interest by PCR. Shrimp alkaline phosphatase (SAP) is used to remove excess dNTPs. A third primer (extension primer) is designed to anneal to a location right next to the mutation site. Depending on the specific mutation introduced and the ddNTP/dNTP mixtures used, either one or two bases are added to the extension primer, producing two extension products from the two templates. The two extension products are then detected and quantified by high-throughput MALDI-TOF MS (**Fig. 1**).

Because DNA quantification by rcPCR is based on its ability to distinguish two DNA sequences differing by a single base, it is also possible to detect and quantify mutants (in the presence of wild-type DNA sequences). A slightly more complicated scheme is needed for assay design, because three competing DNA sequences (standard DNA, wild type DNA, and mutant DNA) are present in the system. However, the experimental procedure remains essentially the same.

In summary, this technique can be used in high-throughput analyses such as:

1. Gene expression analysis. In particular, one can study minor (20–50% down- or upregulation in gene expression), yet biologically significant changes, and allele-specific expression.
2. DNA quantification for analyzing loss of heterozygosity (LOH), trisomies, monosomies, and gene amplification.
3. Mutant DNA detection and quantification in the background of wild-type DNA sequences.

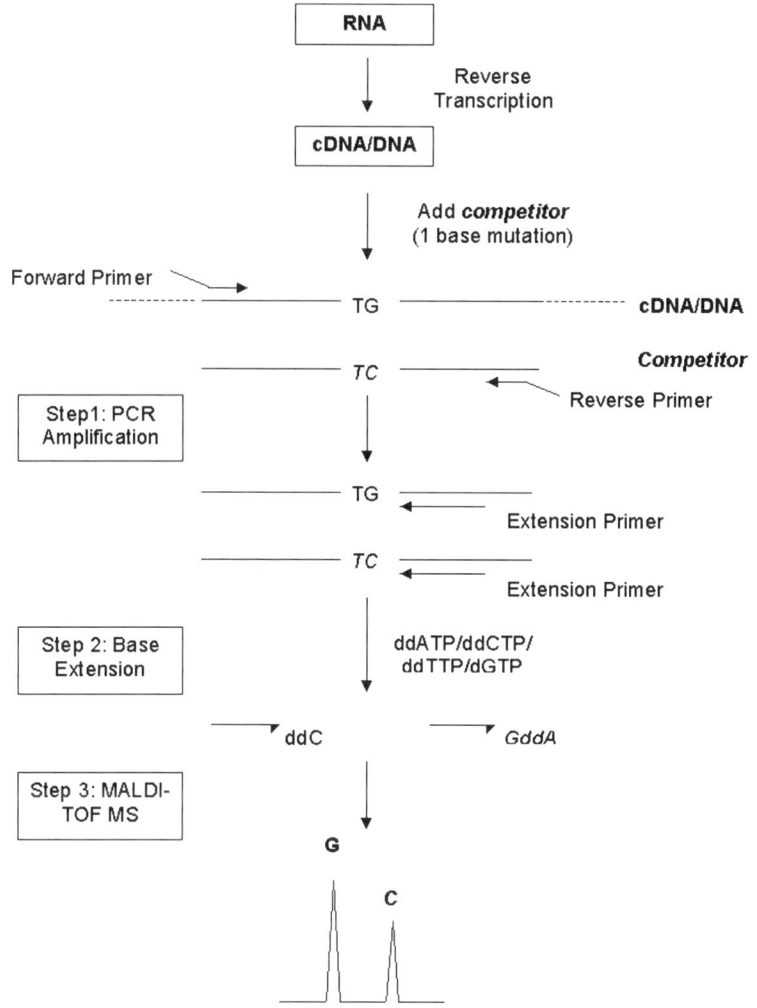

Fig. 1. Flowchart for the real-competitive polymerase chain reaction (rcPCR) approach for DNA quantification. The DNA or cDNA (reverse-transcribed from RNA) sample is spiked with a competing DNA oligonucleotide standard (60–80 bases) with an artificial single-base mutation in the middle of the sequence. These two DNA sequences are co-amplified with PCR in the same reaction. Excess dNTPs are removed by shrimp alkaline phosphatase (not shown). A primer extension reaction is carried out with an extension primer annealing to the region immediately next to the mutation site. Depending on the mutation introduced and the ddNTP/dNTP mixture used, either one or two bases are added to the extension primer. The extension products are detected and quantified by matrix-assisted laser desorption/ionization time-of-flight mass spectrometry.

4. Other DNA/RNA quantifications requiring absolute measurements and/or single base specificity.

2. Materials

1. cDNA or DNA samples.
2. Water (DNase and RNase free, 0.1 μm filtered, Sigma, W4502).
3. Hot Start *Taq* Polymerase (Qiagen).
4. Hot Start *Taq* Buffer (Qiagen).
5. $MgCl_2$.
6 SAP (Sequenom).
7. ThermoSequenase (Sequenom).
8. dNTP mixture (GIBCO), for PCR, 25 m*M* each.
9. ddNTP/dNTP mixtures (Sequenom). Typically, three different ddNTPs and one dNTP are mixed together at equal molar concentrations.
10. PCR and extension primers (IDT Technologies).
11. Oligonucleotide standard (60–80 bases), polyacrylamide gel electrophoresis (PAGE)-purified, concentration determined by absorbance at 260 nm, stored at 10 μ*M* in –80°C.
12. hME buffer (Sequenom).
13. 384- or 96-well microplate (Marsh Biomedical Products) and plate seal (Applied Biosystems, cat. no. 4306311).
14. 96-channel Multimek auto-pipet (Beckman Coulter) for high-throughput analysis or a 12-channel electronic pipet (0.5–10 μL, cat. no. E12-10, Rainin) for low- and medium-throughput analysis.
15. 96-Well SpectroCLEAN plate, SpectroCLEAN resin and scraper (Sequenom).
16. 384-format SpectroCHIP prespotted with 3-hydroxypicolinic acid (Sequenom).
17. SpectroPOINT nanodispenser (Sequenom).
18. A modified Biflex (or Autoflex) MALDI-TOF mass spectrometer with Quantitative Gene Analysis (formerly called Allelotyping) software (Sequenom).

3. Methods

The methods described here include (1) rcPCR assay design and (2) rcPCR protocol including competitive PCR, SAP treatment, primer extension reaction, post-PCR sample processing and nanodispensing, and MALDI-TOF MS. Using the commercially available MassARRAY (Sequenom) system, which is bundled with all necessary softwares for the most efficient and accurate analyses, is highly recommended. Expert users may also be able to replicate the applications in any MALDI-TOF MS-based platforms.

3.1. rcPCR Assay Design

3.1.1. rcPCR for DNA Quantification

This section provides a generic design for quantifying any genomic DNA sequence (i.e., to quantify the presence of an X-chromosome sequence). The main task of the assay design is to identify a short sequence of approx 60–80 bp within the DNA sequence of interest that satisfies the following criteria (*see* **Note 1**).

1. The target sequence should preferably not have a high GC content so that it is easily amplifiable by PCR.
2. Two PCR primers (forward and reverse primers) that anneal to the two ends of the sequence. The PCR primers are typically 20 bases long and have a melting temperature (Tm) above 58°C. A 10-base tag (5'-ACGTTGGATG-3') is added to the 5' end of the PCR primers so that these primers will not interfere in mass spectra.
3. A G/C (or C/G) mutation can be introduced approximately in the middle of the sequence.
4. Directly adjacent to the mutation site (either on the forward or reverse strand), an extension primer (*see* **Note 2**) of 16–25 bases (approx 4800 to 7500 Da) with a Tm of 55°C or higher can be identified.

3.1.2. rcPCR for DNA Mutation Detection and Quantification

This design applies to situations in which both a wild-type and a mutant DNA sequence may be present and a researcher wants to detect and quantify both DNA sequences (allele-specific quantification). Assay designs are slightly more complicated here, because three extension products (from the DNA standard, the mutant DNA, and the wild-type DNA) may be produced. In addition, one or more pausing products may also be present. In **Fig. 1** , if only G is added to the extension primer (instead of GddA), we will obtain a pausing product (**Fig. 2**). The pausing product will interfere in the mass spectrum if its molecular weight is close to one of the normal extension products. Under **Subheading 3.1.1.**, because the G/C (or C/G) mutation is chosen, there will be no pausing product. However, natural mutations are not always G/C (or C/G) mutations. There are two possible options for DNA mutation detection and quantification.

1. Only the relative abundance between a wild type DNA and a mutant DNA is quantified. The assay design is very similar to that under **Subheading 3.1.1.**, except that we do not add an oligonucleotide standard with an artificial mutation. Because a natural mutation is not necessarily a G/C mutation, extra care must be taken to ensure that the pausing product does not have molecular weight that is very similar to the extension products.
2. The absolute concentrations of a wild-type and a mutant DNA are quantified. An example is given in **Fig. 2**. The wild-type DNA has a C whereas the mutant DNA has a G at the polymorphic site. The oligonucleotide standard (competitor) is designed to have a T/A mutation next to the G/C polymorphism. These three DNA sequences are co-amplified by PCR. For base extension reaction, a ddC/ddG/ddT/dA mixture is used to produce three extension products and a possible pausing product. The synthetic mutation in the oligonucleotide standard is

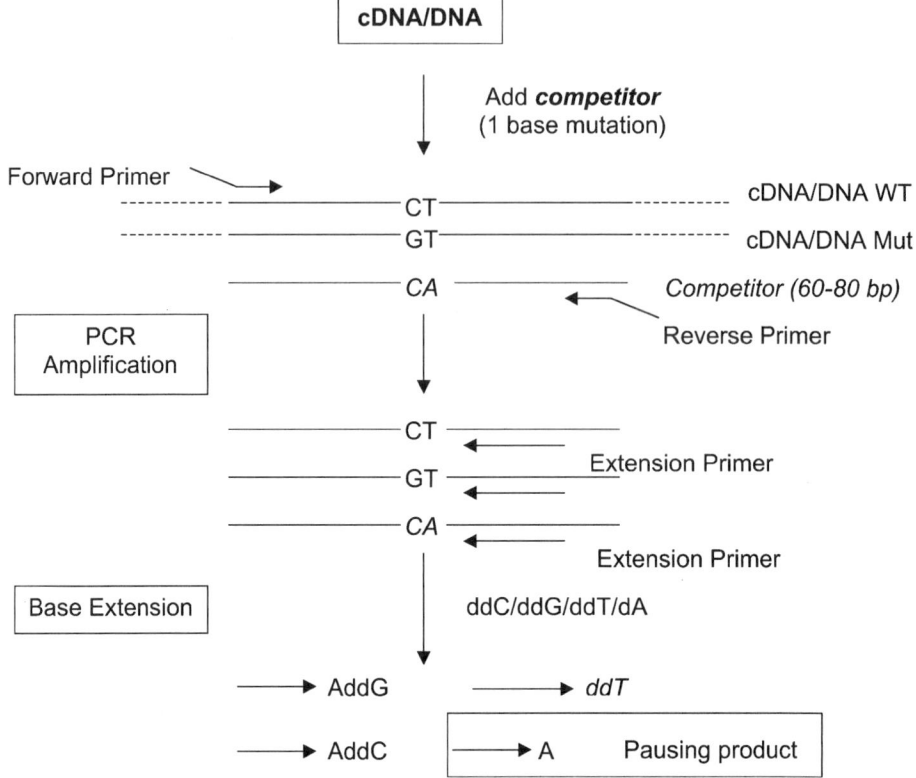

Fig. 2. Schematic representation of allele-specific quantification. When both the wild-type and the mutation DNA are present, the competitor is designed to have a synthetic mutation next to the natural mutation. Three extension products from the wild-type DNA, the mutant DNA, and the competitor are designed to have different molecular weights, and thus can be detected by matrix-assisted laser desorption/ionization time-of-flight mass spectrometry.

designed such that the molecular weight values of the aforementioned four products (**Fig. 2**) differ by at least 25 Da from each other.

3.1.3. rcPCR for cDNA Quantification and cDNA Mutation Detection and Quantification

Generally, assay designs for cDNA quantification (or gene expression analysis) are very similar to DNA quantifications. To avoid potential genomic DNA contamination, it is better to design a PCR amplicon spanning a long intron. If this is not possible, a negative control experiment using nonreverse-transcribed RNA samples should be performed.

3.1.4. Mutant DNA Detection in the Presence of a Large Excess of Wild-Type DNA

On at least two occasions, the robust detection of mutant DNA in the presence of a large excess of wild type DNA may be of clinical and biological significance. One involves cancer mutant DNA in plasma *(12)*. The other involves fetal-specific DNA in maternal plasma *(13)*. In both cases, it is extremely difficult to detect single-base mutations when 99% or more of the DNA is normal. A method called single allele base extension reaction (SABER) was recently developed *(14)*. The SABER technique uses a termination mixture (in the primer extension) that will only extend the mutant DNA (e.g., only ddATP is added for the primer extension in **Fig. 3**). This selective extension step results in robust detection of mutant DNA with potentially 100% specificity and sensitivity by effectively eliminating the background signal from the normal DNA (**Fig. 3**). The SABER technique may be useful for early cancer diagnosis and noninvasive diagnosis of genetic diseases such as β-thalassemia and cystic fibrosis *(14)*.

3.2. rcPCR Experimental Procedure

For gene expression analysis, it is assumed that cDNA samples have already been obtained through reverse transcription using either random hexamers, poly-T primers, or gene-specific primers (*see* **Note 3**).

3.2.1. Polymerase Chain Reaction

A protocol for carrying out PCR reactions in a 384-well microplate is provided in this section. Generally, one can scale up or down.

1. Prepare a PCR cocktail as specified in **Table 1**. For the SABER method, no competitor is needed; thus, 1 µL DNA is used.
2. Add 5 µL of PCR cocktail into each well of a 384-well microplate. Seal the plate and centrifuge it for 3 min at 600g.
3. Perform PCR as follows:
   a. 95°C, 15 min.
   b. 95°C, 20 s.
   c. 56°C, 30 s.
   d. 72°C, 1 min.
   e. Perform **step b** 44 times.
   f. 72°C, 3 min.
   g. 4°C, forever.

3.2.2. Shrimp Alkaline Phosphatase Treatment

This step is used to neutralize remaining dNTPs from the PCR reactions into dNDPs so that they will not interfere with the subsequent primer extension reaction.

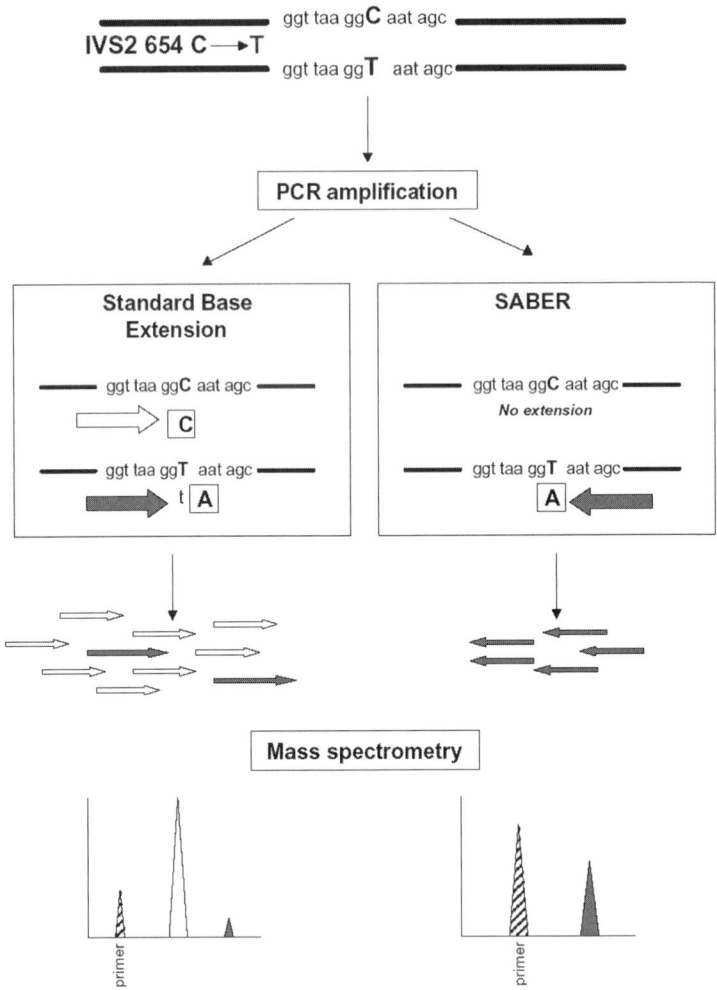

Fig. 3. Schematic illustration of the single allele base extension reaction (SABER) and standard MassARRAY assays. Maternal plasma detection of the paternally inherited fetal specific β-thalassemia mutation, IVS2 654 C>T, is presented as an illustrative example. Maternal plasma is first amplified by polymerase chain reaction (PCR). The PCR products are subjected to base extension by the standard and SABER protocols. The standard protocol involves the base extension of both the mutant fetal allele (the T allele) and the background allele (the C allele), whereas the SABER method only extends the fetal specific mutant allele, because only ddATP is supplied. The base extension reactions are terminated by ddNTPs, indicated in square boxes. The extension products of the standard protocol include a predominance of the nonmutant allele (unfilled block arrows), with a small fraction of the fetal specific mutant allele

Table 1
Preparation of Polymerase Chain Reaction Cocktail

reagent	Final concentration	Volume for one reaction	Volume for one 384-plate
Water (*see* **Subheading 2.**)	N/A	2.24 µL	1120 µL 10X
Hot Start *Taq* PCR buffer, containing 15 m*M* MgCl$_2$	1X 1.5 m*M* MgCl$_2$	0.50 µL	250 µL
25 m*M* MgCl$_2$	1 m*M* MgCl$_2$	0.20 µL	100 µL
dNTP mix, 25 m*M* each	200 µ*M* each	0.04 µL	20 µL
Hot Start *Taq* Polymerase, 5 U/ µL	0.1 U/reaction	0.02 µL	10 µL
Forward and reverse PCR primers (1 µ*M* each)	200 n*M*	1.00 µL	500 µL
Oligonucleotide standard (*see* **Note 4**)		0.50 µL	250 µL
DNA/cDNA (5 ng/µL; *see* **Note 5**)	2.5 ng/reaction	0.50 µL	250 µL
Total volume		5.00 µL	2500 µL

Table 2
Preparation of Shrimp Alkaline Phosphatase Reaction Solution

Reagent	Volume for one reaction	Volume for one 384-well plate
Water	1.53 µL	881.3 µL
hME buffer	0.17 µL	97.9 µL
SAP	0.30 µL	172.8 µL
Total volume	2.00 µL	1152.0 µL

1. Prepare the SAP reaction solution as described in **Table 2**. Make sure the solution is mixed well because this solution can be viscous as a result of the high concentration of glycerol.
2. Add 2 µL of the solution from **step 1** to each well of the 384-plate from the PCR reaction. Each well now contains 7 µL solution.

Fig. 3. (*continued from opposite page*) (filled block arrows). The low abundance of the fetal allele (filled peak) is overshadowed by the nonmutant allele (unfilled peak) in the mass spectrum. As SABER involves the extension of only the mutant allele, the latter's presence (filled peak) can be robustly identified from the mass spectrum. The striped peaks represent the unextended primer.

Table 3
Preparation of Primer Extension Cocktail

Reagent	Stock concentration	Final concentration	Volume for one reaction	Volume for one 384-plate
Water	N/A	N/A	1.674 µL	915.70 µL
ddNTP/dNTP mix	10X Buffer with 2.25 m*M* d/ddNTPs each	1X Buffer with 50 µ*M* d/ddNTPs each	0.200 µL	105.98 µL
Extension primer	100 µ*M*	1.2 µ*M*	0.108 µL	28.62 µL
ThermoSequenase	32 U/µL	0.064 U/µL	0.018 µL	9.54 µL
	Total volume		2.000 µL	1059.84 µL

3. Seal the plate and centrifuge it at 600*g* for 3 min.
4. Thermocycle the sample plate as follows:
    a. 37°C, 20 min.
    b. 85°C, 5 min.
    c. 4°C, forever.

3.2.3. Primer Extension Reaction

In this step, an extension primer is annealed to the region immediately upstream of the mutation site. The ThermoSequenase, three different ddNTPs, and one dNTP are used to extend the primer for one or two bases (terminated by ddNTP). This step produces two extension products (typically 18 to 25 bases long) with different molecular weights. In SABER, only one ddNTP specific for the mutant DNA is used *(14)*.

1. Prepare primer extension cocktail as described in **Table 3**. Make sure the solution is mixed well before aliquoting.
2. Add 2 µL of the solution from **step 1** to each well of the 384-plate from the SAP reaction. Each well now contains 9 µL solution.
3. Seal the plate and centrifuge at 600*g* for 3 min.
4. Thermocycle the sample plate as follows:
    a. 94°C, 2 min.
    b. 94°C, 5 s.
    c. 52°C, 5 s.
    d. 72°C, 5 s.
    e. Perform **step b** 39 times.
    f. 4°C, forever.

3.2.4. Cation Removal

This step is used to remove cations in the solution, because cations might interfere in mass spectra. For high-throughput analyses, please refer to "Cleaning up the hME Reaction Products" in the *MassARRAY Liquid Handler User's Guide* (Sequenom) (all Sequenom manuals mentioned in this chapter come with the Sequenom MALDI-TOF MS system).

1. Place sufficient SpectroCLEAN resin onto a 96-well SpectroCLEAN plate, and then use a scraper to spread the resin into the wells of the SpectroCLEAN plate. Next, scrape off the excess resin.
2. Place a clean, 96-well microplate upside-down over the SpectroCLEAN plate and align all the wells. Flip the two plates over so that the resin is transferred from the SpectroCLEAN plate to the 96-well plate. Tap on the SpectroCLEAN plate, if necessary, to fully release the resin.
3. Add 70 µL water to each well of the 96-plate with resin and thoroughly mix with the resin.
4. Add 16 µL resin/water solution to each well of the 384-plate from the primer extension reaction. Seal the plate and place on a rotator for at least 5 min at room temperature to ensure that the resin and the solution are mixed thoroughly.
5. Centrifuge the plate for 3 min at 600g.

3.2.5. SpectroCHIP Spotting

This step transfers approx 10 nL of the final reaction solution after cation removal onto a 384-format SpectroCHIP. For the most accurate results, it is recommended that a 384-plate be spotted onto 2–4 384-format SpectroCHIPs. Please refer to "Dispensing MassEXTEND Reaction Products onto Spectro-CHIPs" in the *MassArray Nanodispenser User's Guide* (Sequenom) for instructions.

3.2.6. Matrix-Assisted Laser Desorption/Ionization Time-of-Flight Mass Spectrometry

The Quantitative Gene Analysis software is bundled with the MALDI-TOF MS as a complete package. The software performs data acquisitions, mass spectrometric peak identifications, noise normalizations, and peak area analyses. Please refer to the "SpectroACQUIRE" chapter in the *MassARRAY Typer User's Guide* for MALDI-TOF MS operations.

4. Notes

1. The AssayDesigner software (Sequenom) is tailored for designing such assays. In practice, a user only needs to create an artificial mutation (C/G or G/C) in the middle of the DNA sequence of interest and input this information into the software. The AssayDesigner can also handle multiplex assays, ensuring that the molecular weight values of all primers (and the extension products derived from them) involved do not interfere with each other in a mass spectrum.
2. The quality of extension primers is crucial for accurate quantifications. Specifically, the (n-1), (n-2), and other by-products resulting from incomplete oligo-

nucleotide synthesis should only be present in a small proportion (preferably less than 10% of the desired oligonucleotide).

3. Using random hexamers to generate cDNA is recommended, because different genes can be quantified from the same cDNA sample. Because amplicon size is only 60–80 bp in the rcPCR approach, a long cDNA sequence is not needed. In addition, it is possible that more even reverse-transcription efficiency across different regions of the whole transcript can be achieved with random hexamers compared with poly-T and gene-specific primers.

4. If the approximate concentration of the DNA/cDNA sequence is not known, using three different dilutions of a standard (in a series of 100-fold dilutions) for a preliminary analysis is recommended. For most applications that do not require extremely accurate quantifications, this step is sufficient. For experiments in which a small change in DNA/cDNA concentration is expected, it is desirable to carry out a second experiment in which oligonucleotide standard is used at a concentration close to the value of that in the first experiment; this gives the most accurate quantification, with a coefficient of variation (CV) typically less than 10% based on four replicates.

5. For gene expression analysis, the amount of cDNA needed is dependent on the expression level of the specific gene being studied. For human gene expression analysis, 1 ng cDNA (defined as cDNA reverse-transcribed from 1 ng total RNA) is sufficient for most genes. We have obtained single copy sensitivity for DNA detections and as few as five cDNA copies for quantitative analysis without any optimization *(10)*. For human genomic DNA, 2.5 ng DNA is typically used. A premix of DNA/cDNA and oligonucleotide standard is desirable. The most critical step in rcPCR is accurate pipetting when making the oligonucleotide standard dilutions and mixing the standard with the DNA/cDNA samples.

Acknowledgments

I want to thank Dr. Charles R. Cantor (Boston University) for his guidance in developing the rcPCR technique. The rcPCR work was supported by a grant from Sequenom to Boston University. The author was also previously supported by Boston University and is currently supported by the Tung Wah Group of Hospitals and by the Research Fund for the Control of Infectious Disease (RFCID) from the Health, Welfare and Food Bureau of the Hong Kong SAR Government.

References

1. Karas, M. and Hillenkamp, F. (1988) Laser desorption ionization of proteins with molecular masses exceeding 10,000 daltons. *Anal. Chem.* **60,** 2299–2301.
2. Tang, K., Shahgholi, M., Garcia, B. A., et al. (2002) Improvement in the apparent mass resolution of oligonucleotides by using 12C/14N-enriched samples. *Anal. Chem.* **74,** 226–231.

3. Tang, K., Fu, D. J., Julien, D., Braun, A., Cantor, C. R., and Koster, H. (1999) Chip-based genotyping by mass spectrometry. *Proc. Natl. Acad. Sci. USA* **96,** 10,016–10,020.

4. Krause, E., Wenschuh, H., and Jungblut, P. R. (1999) The dominance of arginine-containing peptides in MALDI-derived tryptic mass fingerprints of proteins. *Anal. Chem.* **71,** 4160–4165.

5. Koster, H., Tang, K., Fu, D. J., et al. (1996) A strategy for rapid and efficient DNA sequencing by mass spectrometry. *Nat. Biotechnol.* **14,** 1123–1128.

6. Van Ausdall, D. A. and Marshall, W. S. (1998) Automated high-throughput mass spectrometric analysis of synthetic oligonucleotides. *Anal. Biochem.* **256,** 220–228.

7. Mann, M., Hendrickson, R. C., and Pandey, A. (2001) Analysis of proteins and proteomes by mass spectrometry. *Annu. Rev. Biochem.* **70,** 437–473.

8 Haff, L. A. and Smirnov, I. P. (1997) Single-nucleotide polymorphism identification assays using a thermostable DNA polymerase and delayed extraction MALDI-TOF mass spectrometry. *Genome Res.* **7,** 378–388.

9. Gygi, S. P., Rist, B., Gerber, S. A., Turecek, F., Gelb, M. H., and Aebersold, R. (1999) Quantitative analysis of complex protein mixtures using isotope-coded affinity tags. *Nat. Biotechnol.* **17,** 994–999.

10. Ding, C. and Cantor, C. R. (2003) A high-throughput gene expression analysis technique using competitive PCR and matrix-assisted laser desorption ionization time-of-flight MS. *Proc. Natl. Acad. Sci. USA* **100,** 3059–3064.

11. Ding, C., Maier, E., Roscher, A. A., Braun, A., and Cantor, C. R. (2004) Simultaneous quantitative and allele-specific expression analysis with real competitive PCR. *BMC Genet.* **5,** 8.

12. Chen, X. Q., Stroun, M., Magnenat, J. L., et al. (1996) Microsatellite alterations in plasma DNA of small cell lung cancer patients. *Nat. Med.* **2,** 1033–1035.

13. Lo, Y. M. D., Corbetta, N., Chamberlain, P. F., et al. (1997) Presence of fetal DNA in maternal plasma and serum. *Lancet* **350,** 485–487.

14. Ding, C., Chiu, R. W., Lau, T. K., et al. (2004) MS analysis of single-nucleotide differences in circulating nucleic acids: Application to noninvasive prenatal diagnosis. *Proc. Natl. Acad. Sci. USA* **101,** 10,762–10,767.

7

Analysis of Polymerase Chain Reaction Products by Denaturing High-Performance Liquid Chromatography

Ching-Wan Lam

Summary

Denaturing high-performance liquid chromatography (DHPLC) analysis is an ion-pair reversed-phase high performance liquid chromatography for performing analytical separations of DNA based on temperature: analysis is performed at a temperature sufficient to partially denature DNA heteroduplexes. The technology detects single-base changes as efficiently as short deletions and insertions. The chance that a mutation cannot be detected is 0.5%.

Key Words: Denaturing HPLC; mutation analysis; heteroduplex; automation.

1. Introduction

Denaturing high-performance liquid chromatography (DHPLC) analysis is an ion-pair reversed-phase high performance liquid chromatography for performing analytical separations of DNA based on temperature: analysis is performed at a temperature sufficient to partially denature (melt) DNA hetero-duplexes *(1)*. The melted heteroduplexes are resolved from the corresponding homoduplexes by ion-pair reversed-phase high performance liquid chromatography. The procedure is referred as temperature-modulated heteroduplex chromatography (TMHC). THMC relies on the physical changes in DNA molecules induced by mismatched heteroduplex formation. Heteroduplex DNA is generated by denaturing and re-annealing a mixed population of reference or "wild-type" sample and "mutant" DNA (**Fig. 1**). The heteroduplex DNA fragments form as a result of the base-pairing of a single-stranded, mutated DNA with a single-stranded, "wild-type" DNA. The two strands will not form hydrogen bonds at the mutation site because the basepairs are mismatched, thus giving the heteroduplex different melting properties than the homoduplex. At a critical temperature, heterduplexes will begin to "melt" or denature, mak-

From: *Methods in Molecular Biology, vol. 336: Clinical Applications of PCR*
Edited by: Y. M. D. Lo, R. W. K. Chiu, and K. C. A. Chan © Humana Press Inc., Totowa, NJ

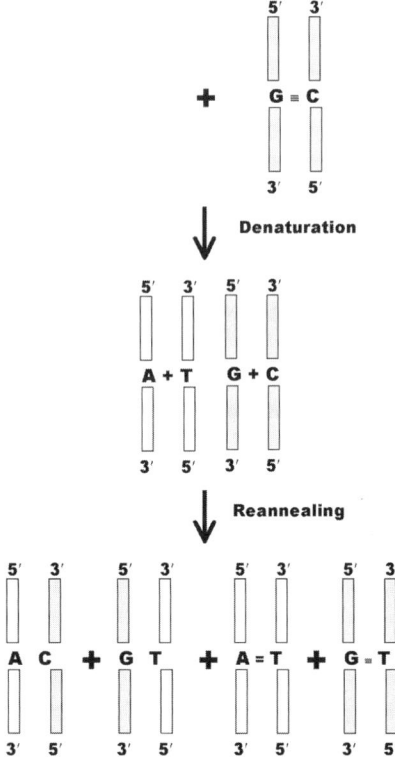

Fig. 1. Schematic presentation of heteroduplex formation through hybridization.

ing part of the heteroduplexes close to the mismatched base single-stranded whereas the homoduplexes remain double-stranded. The percentage of organic mobile phase that disrupts the interactions between DNA fragments and the DNAsep cartridge matrix is lower for heteroduplex DNA strands than for homoduplex DNA strands. Therefore, heteroduplex DNA fragments elute earlier in the gradient, and this signals the presence of a mutation.

DHPLC is fully automated, eliminating time and labor in gel preparation, loading, running, analyzing, and photographic documentation. The polymerase chain reaction (PCR) products are directly loaded without any prior purification as would be required in various enzymatic methods and capillary electrophoresis. The separation efficiency and the optimized time for DNA fragment elution and column equilibrium allow the analysis of up to 200 samples per day on an instrument with a single analytical column.

A number of clinical applications for mutational analysis have been published (2–9), and disease-causing genes that have been screened by DHPLC can be found on the website of Dr. Oefner (http://insertion.stanford.edu/

dhplc_genes2.html). In this chapter, we use the *G6PT1* gene as an example to illustrate the use of DHPLC for prenatal diagnosis of a genetic disease *(7)* (*see* **Note 1**).

Glycogen storage disease Ib (GSD Ib) is an inborn error of metabolism with autosomal recessive inheritance, caused by a deficiency in glucose-6-phosphate translocase (G6PT1). Patients with GSD Ib present with hypoglycemia, hepatomegaly, kidney enlargement, growth retardation, lactic acidemia, hyper-lipidemia, hyperuricemia, neutropenia and impaired neutrophil function, oral and intestinal mucosa ulcerations, and inflammatory bowel disease. The G6PT1 protein has not been purified to homogeneity, and its molecular nature remained unknown until recently.

Gerin et al. *(10)*, using a candidate gene approach, identified mutations of a putative human *G6PT1*, homologous to bacterial transporters of hexose-6-phosphate, in patients with GSD Ib. The gene is composed of nine exons spanning a genomic region of approx 5 kb *(11)*. The liver cDNA encodes a protein of 429 amino acids. We have identified a patient clinically suspected of GSD Ib *(12)*. The patient has all of the clinical features of GSD Ib, including neutropenia. The patient is the only child in the family, and the parents are not consanguineous. DNA sequencing of all nine exons and the exon-intron boundaries revealed a homozygous G>A transition at nucleotide position 1563 of the genomic DNA sequence. The missense mutation is located in exon 3, at codon 149, changing the nonpolar glycine to the charged glutamic acid, i.e., G149E. The father and mother are each heterozygous for this mutation. In this study, we have used DHPLC for prenatal diagnosis of this disease in her second pregnancy.

2. Materials
1. Genomic DNA template.
2. QIAamp Tissue Kit (Qiagen).
3. PCR primers.
4. PCR kit (*see* **Note 2**).
5. Thermal block or water bath.
6. WAVE™ DHPLC analyzer and software (Transgenomic Inc., San Jose, CA).
7. DNASep cartridge packed with C18 alkylated, polystyrene-divinylbenzene polymeric beads.
8. Eluent A: 0.1 *M* triethylammonium acetate (TEAA).
9. Eluent B: 0.1 *M* TEAA and 25% acetonitrile (ACN).

3. Methods
3.1. PCR Amplification
1. Genomic DNA is extracted from cultured chorionic villi using a QIAamp Tissue Kit (Qiagen) according to the manufacturer's instructions.

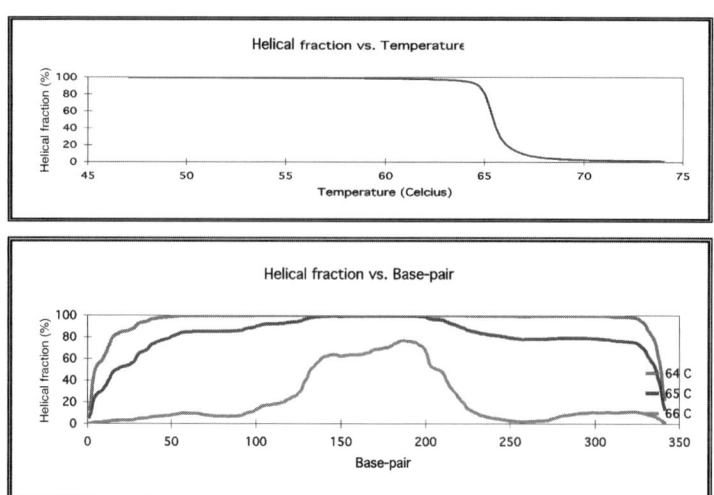

Fig. 2. The melting profile of exon 3 of the *G6PT1* gene. The DNA sequence of the amplicon is: CTGCCCCATCTGACCCCACCCTCAacatgggcagtaggctggacacctac gtgtcgccctctgccccacagtggtttgagccatctcagtttggcacttggtgggccatcctgtcaaccagcatgaacctg gctg**g**agggctgggccctatcctggcaaccatccttgcccagagctacagctggcgcagcacgctggccctatctg gggcactgtggtgtggttgtctccttcctctgtctcctgctcatccacaatgaacctgctgatgttggact ccgcaacctggaccccatgccctctgagggcaagaagggtGAGCCCCCACCCAGACCGACCACT. The uppercase letters are primer sequences. The letter in bold type is the nucleotide position of the mutation in the wild-type sequence, i.e., 1563G>A. The mutation is located at nucleotide position 136 of the amplicon. The sequence is shown in the sense of direction.

2. PCR amplification (*see* **Notes 3** and **4**) is conducted using the primer pairs and conditions described elsewhere *(13)* (**Fig. 2**).

3.2. DHPLC Analysis

DHPLC analysis is typically conducted at temperatures from 51 to 75°C. The WAVE utility software (Transgenomic Inc.) helps determine the correct temperature for mutation scanning based on the sequence of the wild-type DNA. The calculated melting behavior is then used to predict the temperature suitable for mutation scanning of the fragment. Separation of heteroduplex DNA from homoduplex DNA depends on a greater proportion of the nonhelical form in the vicinity of the mismatched bases. To achieve the best possible separation, the region containing the mutation should have a helical fraction between 30 and 99%. For example, for DHPLC analysis of exon 3 of the *G6PT1* gene, the predicted melting temperature is 65°C (**Fig. 2**). The DNASep Cartridge is packed with C18 alkylated, polystyrene-divinylbenzene polymeric

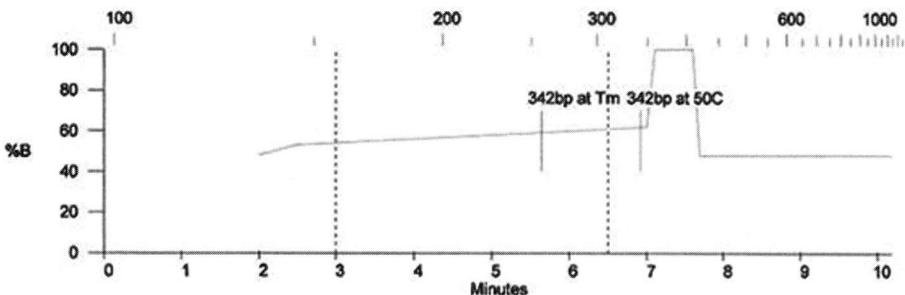

Fig. 3. Chromatographic gradient for mutation detection of exon 3 of the *G6PT1* gene.

beads (average diameter 2.3 μm), allowing for analysis under a wide range in pH (2.0–13.0) and temperature (40–80°C) conditions. A positively charged ion-pairing reagent, TEAA, allows the negatively charged DNA backbone to interact with the hydrophobic DNAsep cartridge matrix.

1. Heat the PCR product to 95°C for 3 min and allow it to cool slowly. For an individual heterozygous for the targeted mutation, the wild-type and mutant DNA hybridize to form a mixture of hetero- and homoduplexes. This approach is modified during the analysis of DNA from individuals carrying two identical mutant alleles (homozygous mutation). The PCR product spanning the homozygous mutation is mixed with wild-type amplified DNA and hybridized. After this treatment, a sample will contain a mixture of hetero- and homoduplexes. In our case, we have a homozygous mutation; therefore, in the first part of the analysis, we analyzed the PCR product amplified from the fetal DNA. In the second part of the analysis, we mix the PCR product amplified from the fetal DNA with that amplified from wild-type DNA. The nucleotide sequence of *G6PT1* in the wild-type sample has been confirmed to be normal by direct DNA sequencing.
2. Eluents A (0.1 *M* TEAA) and B (0.1 *M* TEAA and 25% ACN) are prepared from concentrated TEAA (100 mL) (*see* **Note 5**).
3. Between 5 and 10 μL of crude PCR product is loaded on a DNASep column (*see* **Note 6**).
4. The PCR product is eluted from the column by an acetonitrile gradient in 0.1 mol/L TEAA, pH 7.0, at a constant flow rate of 0.9 mL/min (*see* **Note 7**). The gradient is created by mixing eluents A and B. The recommended gradient for mutation detection is a slope of 2% increase in buffer B per minute (*see* **Note 8**); e.g., the gradient for DHPLC analysis of exon 3 of the *G6PT1* gene is depicted in **Table 1** and **Fig. 3**.
5. Eluted DNA fragments are detected with ultraviolet absorption at wavelength 260 nm by the detector within the DHPLC analyzer (*see* **Note 9**).

Table 1
Mobile Phase Conditions for Gradient Elution

Step	Time	%A	%B	mL/min	At detector
Loading	0.00	52	48	0.9	2.00
Start gradient	0.50	47	53		2.50
Stop gradient	5.00	38	62		7.00
Start clean	5.10	0	100		7.10
Stop clean	5.60	0	100		7.60
Start equilibrate	5.70	52	48		7.70
Stop equilibrate	8.20	52	48		10.20

6. A final increase in organic mobile phase cleans the DNASep cartridge matrix, removing DNA and other contaminants and preparing the DNASep Cartridge for the next analysis (*see* **Notes 10–12**).

3.3. Interpretation

In the sample with only the PCR products amplified from the fetal DNA, a single peak is observed in the chromatogram (**Fig. 4**). In the sample with PCR products amplified from the wild-type DNA mixed with that amplified from the fetal DNA, two peaks are observed (**Fig. 5**). These two findings together indicate that the patient is homozygous for the G149E mutation. The fetus is predicted to be affected with GSD Ib. On direct DNA sequencing, a homozygous 1563G>A is found in the fetal DNA (not shown). This finding confirms the DHPLC results (*see* **Notes 13** and **14**).

4. Notes

1. The website www.MutationDiscovery.com, hosted by Transgenomic, provides researchers access to experimental methods, methods for DHPLC analysis on a specific gene, and publications on DHPLC analysis of the gene. More than 350 human genes have been screened entirely or partly by DHPLC.
2. Use metal-free DNA polymerase buffers, such as *Taq* Gold buffer or Optimase polymerase buffer.
3. Do not use mineral oil for topping the reaction mixture.
4. Use no more than 100 ng DNA for the PCR reaction. Otherwise, the system will quickly be contaminated by injected DNA.
5. Do not use autoclaved water. The autoclaving process can introduce metal and other ionic contaminates, which would bind and cause damages to the column.
6. For quality control of DHPLC analysis, we run control samples before and after each DHPLC analysis. DHPLC users can try to reduce the cost of analysis by replacing the commercial mutation standards with in-house validated controls.

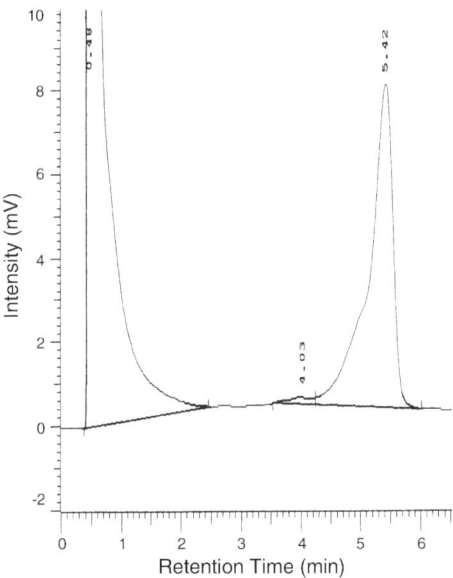

Fig. 4. Chromatogram of polymerase chain reaction products amplified from the fetal DNA. (Reproduced from **ref. 2**, with the permission of John Wiley and Sons Ltd.)

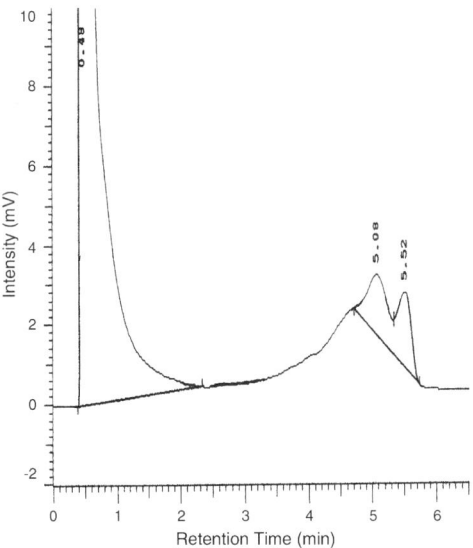

Fig. 5. Chromatogram of a mixture of polymerase chain reaction products amplified from the fetal DNA and the wild-type DNA. (Reproduced from **ref. 2**, with the permission of John Wiley and Sons Ltd.)

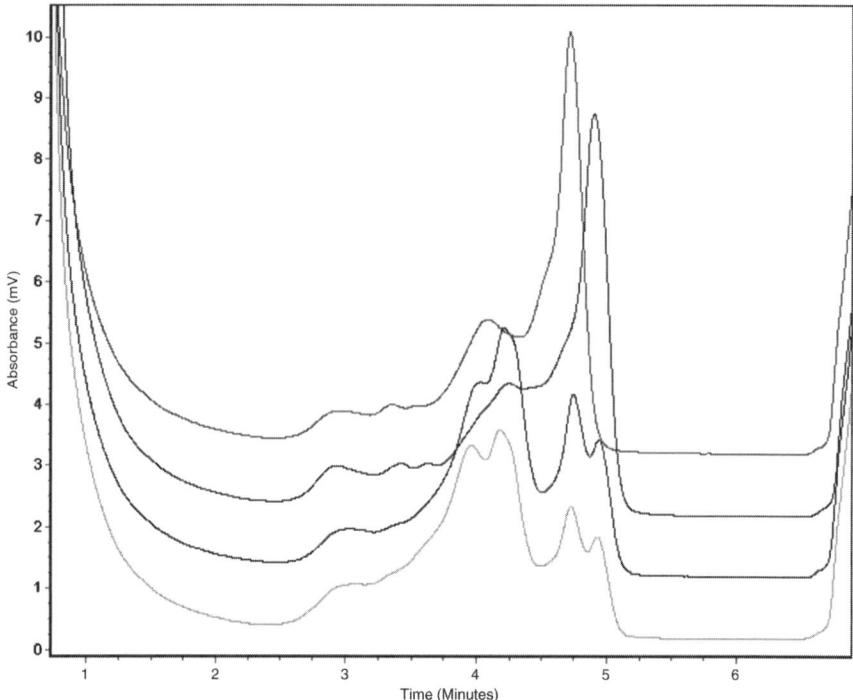

Fig. 6. From top to bottom: patient homozygous for R160Q of the *SUOX* gene, a normal individual; mother of the patient, polymerase chain reaction product mix of the patient and the normal individual.

7. The pressure of the system should be approx 104 psi at 0.05 mL/min flow of 50% A/B and 1560 psi at 0.9 mL/min. The pressure should increase very slowly with the increase in the number of injections. Any drastic increase signals severe clogging of the column or other parts of the system. The column should be reverse hot-washed and the whole system (with column removed) should be thoroughly cleaned with 50% isopropanol.
8. Use a 1.2–1.8% B/min gradient for the separation of larger fragments.
9. Elution profiles of each single injection should be minimally 2 mV.
10. To clean the column, run a Dwash for 15 min at 75°C at the end of each project.
11. To extend the life span of the column when not in use, either keep the column flushed with 50% A/B at 0.05 mL/min, or store the column capped in 75% ACN.
12. When the flow of eluent is to be turned off for an extended period of time, i.e., 3 d or more, remember to flush the separation cartridge with 75% ACN for 15 min, and remove it from the flowpath. Cap the ends of the cartridge tightly with the supplied plugs, and replace with a PEEK union. Prior to returning the cartridge to the system, flush the system with 75% ACN for 10 min; return the

cartridge to the system, and flush the cartridge with 75% ACN for 15 min. Remember to equilibrate the cartridge with 50% A and 50% B at the appropriate flow rates for 30 min.

13. Standard procedures for operating a DHPLC instrument for mutation detection (e.g., normal ranges for the critical parameters of the mutation standards) have been published *(14)*.

14. DHPLC can separate wild-type homoduplexes from mutant homoduplexes for certain mutations (**Fig. 6**).

References

1. Oefner, P. J. and Huber, C.G. (2002) A decade of high-resolution liquid chromatography of nucleic acids on styrene-divinylbenzene copolymers. *J. Chromatogr. B. Analyt. Technol. Biomed. Life Sci.* **782,** 27–55.

2. Lam, C. W., Sin, S.Y., Lau, E., et al. (2000) Prenatal diagnosis of glycogen storage disease type 1b using denaturing high-performance liquid chromatography. *Prenat. Diag.* **20,** 765–768.

3. Lam, C. W., Yeung, W. L., Ko, C. H., et al. (2000) Spectrum of mutations in the MECP2 gene in patients with infantile autism and Rett syndrome. *J. Med. Genet.* **37,** e41.

4. Lam, C. W., Hui, K. N., Poon, P. M. K., et al. (2001) Novel splicing mutation of the PPOX gene (IVS10+1G>A) detected by denaturing high-performance liquid chromatography. *Clin. Chim. Acta.* **305,** 197–200.

5. Lam, C. W., Poon, P. M. K., Tong, S. F., et al. (2001) Novel mutation and polymorphisms of the HMBS gene detected by denaturing HPLC. *Clin. Chem.* **47,** 343–346.

6. Lam, C. W., Ko, C. H., Poon, P. M. K., and Tong, S. F. (2001) Two novel CLN2 gene mutations in a Chinese patient with classical late-infantile neuronal ceroid lipofuscinosis. *Am. J. Med. Genet.* **99,** 161–163.

7. Lam, C. W. (2002) First application of denaturing high-performance liquid chromatography (DHPLC) for prenatal diagnosis of genetic disease. *Prenat. Diagn.* **22,** 79–80.

8. Lam, C. W., Li, C. K., Lai, C. K., et al. (2002) DNA-based diagnosis of isolated sulphite oxidase deficiency by denaturing high-performance liquid chromatography. *Mol. Genet. Metab.* **75,** 91–95.

9. Lam, C. W., Leung, C. Y., Lee, K. C., et al. (2002) Novel mutations in the PATCHED gene in basal cell nevus syndrome. *Mol. Genet. Metab.* **76,** 57–61.

10. Gerin, I., Veiga-da-Cunha, M., Achouri, Y., Collet, J. F., and Van Schaftingen, E. (1997) Sequence of a putative glucose-6-phosphatase translocase, mutated in glycogen storage disease type Ib. *FEBS Lett.* **419,** 235–238.

11. Ihara, K, Kuromaru, R, and Hara, T. (1998) Genomic structure of the human glucose 6-phosphate translocate gene and novel mutations in the gene of a Japanese patient with glycogen storage disease type Ib. *Hum. Genet.* **103,** 493–496.

12. Lam, C. W., Tong, S. F., Lam, Y. Y., Chan, B. Y., Ma, C. H., and Lim, P. L.(1999) Identification of a novel missense mutation (G149E) in glucose-6-phosphate

translocase gene in a Chinese family with glycogen storage disease Ib. *Hum. Mut.* **13,** 507.

13. Galli, L., Orrico, A., Marcolongo, P., et al. (1999) Mutations in the glucose-6-phosphate transporter (G6PT) gene in patients with glycogen storage diseases type Ib and 1c. *FEBS Lett.* **459,** 255–258.

14. Schollen, E., Dequeker, E., McQuaid, S., et al. (2005) Diagnostic DHPLC Quality Assurance (DDQA): a collaborative approach to the generation of validated and standardized methods for DHPLC-based mutation screening in clinical genetics laboratories. *Hum. Mutat.* **25,** 583–592.

8

Use of Real-Time Polymerase Chain Reaction for the Detection of Fetal Aneuploidies

Bernhard Zimmermann, Lisa Levett, Wolfgang Holzgreve, and Sinuhe Hahn

Summary

With the advent of real-time polymerase chain reaction (PCR), it is now possible to measure nucleic acid concentrations with an accuracy that was not possible only a few years ago. Examples are the analysis of gene expression or gene duplications/losses, where twofold differences in nucleic acid concentration have routinely been determined with almost 100% accuracy. As our primary interest is in prenatal diagnosis, we have investigated whether real-time PCR could be used for the diagnosis of chromosomal anomalies, in particular the aneuploidies such as trisomy 21, where the difference in copy number is only 50%. The feasibility of such an approach was first tested in a pilot study, in which we were able to demonstrate that trisomy 21 samples could be detected with 100% specificity. We have recently modified this test in order to permit the simultaneous analysis of trisomies 18 and 21, and have in a large scale analysis demonstrated that our approach can be used for the highly reproducible and robust detection of only 1.5-fold differences in gene copy number.

Key Words: Multiplex real-time PCR; relative quantification; prenatal diagnosis; amniotic fluid; amniocentesis; trisomy 21; Down syndrome; trisomy 18; karyotype; aneuploidy; genetics; DNA analysis; gene copy number; molecular probes; PCR primers; PCR kinetics.

1. Introduction

1.1. Background

The detection of gross chromosomal abnormalities is a major focus of prenatal diagnostics. The most common cytogenetic anomaly in live births is trisomy 21, also known as Down syndrome. Other fetal aneuploidies frequently detected involve chromosomes 13, 16, 18, and both sex chromosomes (X and Y). Currently, prenatal diagnosis of genetic anomalies relies on invasive

From: *Methods in Molecular Biology, vol. 336: Clinical Applications of PCR*
Edited by: Y. M. D. Lo, R. W. K. Chiu, and K. C. A. Chan © Humana Press Inc., Totowa, NJ

procedures such as amniocentesis and chorionic villus sampling (CVS), from which the full fetal karyotype is usually determined using cultured cells. The 2-wk period needed for cultivation and subsequent analysis has proven to be associated with considerable parental anxiety and medical problems in those situations requiring therapeutic intervention.

In order to address these needs, more rapid methods for the prenatal diagnosis of fetal chromosomal aneuploidies have recently been developed and implemented *(1)*. The first of these to be commercially introduced was multi-color fluorescence *in situ* hybridization (FISH) for uncultured cells. Although this method is very reliable and has proven in large-scale studies to be very accurate, it is a time- and labor-intensive procedure. Furthermore, as this method requires intact cells, it can only be used on fresh or specially stored samples *(2–4)*. The next method that has seen widespread clinical application, particularly in the United Kingdom, is quantitative fluorescent polymerase chain reaction (PCR) analysis of short tandem repeats (STRs) *(5–8)*. This method has also proven itself to be rapid and reliable, once the initial problems with polymerase stuttering and amplification failure of the highly repetitive loci had been overcome.

An important point to bear in mind regarding both of these rapid diagnostic tests is that they currently only permit a result regarding the most common fetal aneuploidies (chromosomes X, Y, 13, 18, and 21). Hence, it is still necessary to resort to the normal 2-wk cell culture-based analysis in order to obtain a full karyotype. Maternal blood contamination of the amniotic fluid or chorionic villus sample, although infrequent (affecting less than 2% of samples collected), can interfere with either method of analysis, as the results cannot be distinguished from cases of fetal mosaicism.

The recent development of real-time PCR has rapidly emerged as a powerful tool for the accurate and precise determination of template copy numbers, and has found widespread applicability in the analysis of gene expression and cell-free DNA in body fluids as well as the measurement of gene duplications or deletions in cancer research *(9–13)*. The precision of current assays and technology, however, was not thought to permit analyses of less than twofold differences of target template concentrations. As our primary interest is in the development of new methods for prenatal diagnosis, we were interested in whether this new technology could be used for the determination of fetal trisomies. This would, however, entail the resolution of less than twofold increments in target gene copy number, as in these instances there is only a 50% increase present in the amount of a particular chromosome.

Our reason for choosing a real-time "TaqMan" PCR approach is that we have considerable experience with the methodology *(14–17)*, especially in the quantitation of trace quantities of cell-free fetal DNA in the maternal circulation. To test the feasibility of this approach, we first performed a small-scale study involving trisomy 21, as this is the most common aneuploidy in live births. In our study, we compared the gene dosage of a sequence on the Down's critical region on chromosome 21 with a control locus on chromosome 12. Our analysis indicated that by using such an approach of comparative quantitation it was indeed possible to discern between karyotypically normal samples and trisomic ones, provided that certain criteria concerning quantity and quality of DNA and replicate reaction uniformity were met *(18)*. After successful completion of this pilot study, we have now extended this method for the simultaneous detection of trisomy 18 and trisomy 21. In this analysis, the two chromosomes being interrogated are quantified relative to one another.

Future improvements would include the development of a test in which the other common fetal aneuploidies (chromosomes 13, 16, X, Y) would be analyzed in a matrix-type assay. By using such a matrix approach, whereby the dosage of each chromosome is measured relative to the other chromosomes, an automatic system of "checks and balances" would be implemented, thereby helping to reduce the potential error rate considerably.

The use of the real-time PCR assay need not be restricted solely to the detection of aneuploidies, but can also be applied for the discernment of translocations and for the distinction of these from chromosomal trisomies. This can be achieved by amplifying sequences critical to the region translocated, such as those located in the Down's region of chromosome 21, which should enable the detection of those unbalanced Robertsonian translocations that occur in approx 4% of Down syndrome cases. It must be noted, however, that the design of the assay does not permit one to detect triploidies, because the chromosomal balance is equal to that of a normal karyotype.

Advantages of the real-time PCR test for prenatal diagnosis are that it is insensitive to maternal blood contamination of samples where the maternal DNA is present at a small percentage (about 20% or less of the total DNA in the sample). On the other hand, this test will not be able to detect low-level chimerism. Akin to the PCR amplification of STRs, the real-time PCR test also can be used on a wide variety of sample materials and does not require fresh or specially stored cells, because it only requires genomic DNA. Furthermore, the assay is readily amenable to automation, and by making use of the current real-time PCR 96- or 384-well formats, it also facilitates high throughput. In addition, by making use of a closed system, whereby the samples are analyzed directly in the PCR reaction vessel during the amplification and do not need to be opened for analysis as is the case for the analysis of STRs, the assay is also less prone to contamination.

1.2. The Principle of Relative Quantification by Real-Time PCR in Relation to the Determination of Chromosome Ploidy

Real-time PCR is based on the detection and quantitation of a fluorescent signal that is directly proportional to the PCR product being generated during each cycle of the PCR. In our assay, the fluorescence is generated by the 5' nuclease method: a so-called TaqMan probe hybridizes specifically to one strand of the DNA sequence between the two primers. The probe is labeled with a fluorescent detector dye at its 5' end and with a quencher dye at its 3' end. The annealing temperature of the probe is 10°C higher than the annealing temperature (T_A) of the PCR reaction, which is the melting temperature of the primers. This ensures that in each cycle of the reaction, a probe binds to every target sequence before the primers anneal. When the primer binds to the target, the *Taq* polymerase immediately starts extending the primer and by its nuclease activity cleaves the 5' end of the probe. This nucleolytic activity separates the detector dye from the quencher dye, thereby permitting the detector dye to emit a characteristic fluorescence signal when excited by an appropriate light source. As the displacement of each probe molecule is the result of a single template amplification, the amount of fluorescent signal measured is therefore directly proportional to the number of probes cleaved and the amount of PCR product synthesized.

For the correct analysis of such real-time PCR assays, several parameters have been devised. The first of these is the C_T-value, or threshold cycle, defined as the cycle number at which point the amplification curve crosses the threshold line in semi-log view of the amplification plot (**Fig. 1**). Although the C_T-value is the chief factor used in most real-time PCR analyses, in our experience another important parameter, especially for the analysis of discrete template increments, is the normalized final fluorescence, ΔR_n. This ΔR_n-value, a measure of the accumulation of specific fluorescence, is indicative of the amplification efficiency and initial template concentration (**Fig. 1**).

As our real-time PCR test for fetal aneuploidy entails the simultaneous amplification of two chromosomal loci (e.g., chromosome 21 vs chromosome 12 or chromosome 21 vs chromosome 18) in a multiplex reaction, their product formation is detected by two different fluorescent dyes. As each of these reporter dyes has a discretely different emission spectrum, and as the amount of each dye measured is proportional to the relevant target template, this enables the relative quantification of both chromosomes. This can be calculated by using the ΔC_T, which is the difference between the C_T-values of the first (e.g., chromosome 18, VIC dye) and the second (e.g. chromosome 21, FAM dye) amplified sequences in one reaction well (**Fig. 2**):

$$\Delta C_T = C_T \text{ (target A)} - C_T \text{ (target B)}$$

$$= C_T \text{ (VIC, Chromosome 18)} - C_T \text{ (FAM, Chromosome 21)}$$

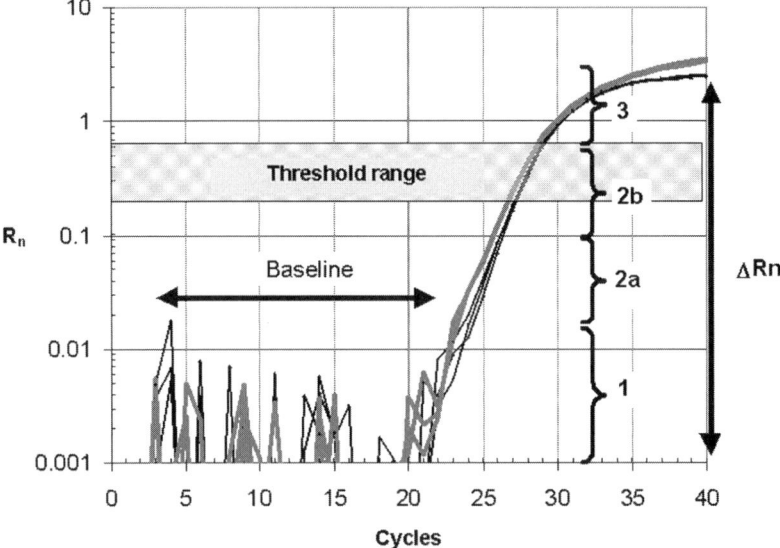

Fig. 1. Amplification plot of the real-time polymerase chain reaction data. (1) Phase of observed background fluorescence. (2) Observable exponential phase. (2a) Measurements are close to the detection limit. As a result of the high contribution of the background fluorescence to the total fluorescence measured, replicate curves can deviate (as seen in the black curves). (2b) Influence of background fluorescence minimal. (3) Linear and plateau phases of the amplification with decreasing amplification efficiencies.

Displayed are triplicate amplifications of a sample with trisomy 21. The grey curves represent data from chromosome 21 (FAM dye), the black curves from chromosome 18 (VIC dye).

In order for this analysis to be accurate, it is important that the co-amplification of two sequences occur in the same reaction vessel. This scenario will guarantee that no well-to-well variation occurs between the two amplified targets *(19)*. In this regard, it is important to realize that the analysis cannot be carried out in a monoplex manner. As with any small well-to-well deviations resulting from minute pipetting errors, alterations in polymerase activity, temperature, or illumination gradients, as well as unequal reagent depletion, will lead to an inaccurate assessment of the target chromosome ploidy. Similarly, it is not possible to determine the ploidy of a sample by referring to a standard curve, as this analysis will be misled by the same factors affecting monoplex analyses, a facet which is not alleviated by post-run data analysis *(22–25)*.

In order to balance small fluctuations and ensure that the amplification of both target sequences has proceeded with similar efficiency, we have devised the following analytical aid: instead of a solitary C_T-value, three to four points

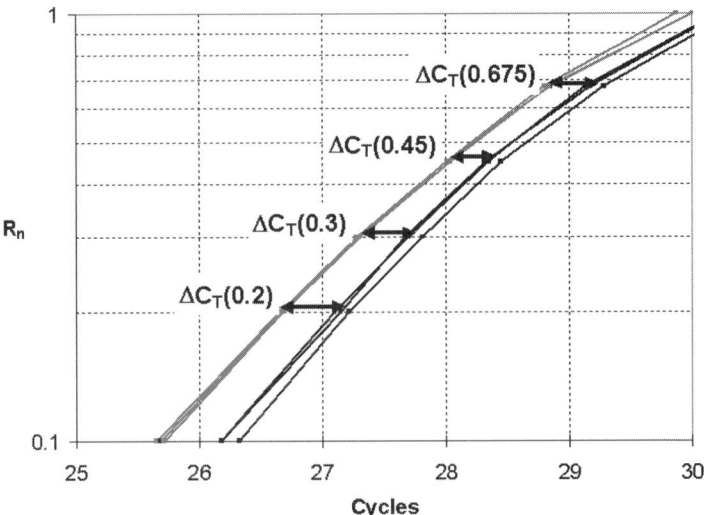

Fig. 2. Change in threshold cycle (ΔC_T) in the amplification plot (magnification of Fig. 1). The ΔC_Ts of the four thresholds, indicated by double-headed arrows.

are chosen along the linear phase of the amplification plot (**Fig. 2**). Those samples in which a deviation is found to occur at these points are either discarded or reanalyzed. If this provision has been made, then the fluorescent signals from both amplifications will be detected simultaneously if the sample is karyotypically normal, i.e., both chromosomes being interrogated have the same copy number (**Fig. 3**). In case of a trisomy, the chromosome present at three copies per cell will be detected at a lower (earlier) C_T-value than the other.

The measured differences in C_T (the ΔC_T) between the two target chromosomes can be converted into a ratio *(20)*:

$$\text{target A/target B} = 2^{-(\Delta C_T)} = \text{Chromosome 18/Chromosome 21}$$

In theory, the difference in threshold cycle number (ΔC_T) for a normal (and for a triploid) karyotype will be 0 cycle; for a trisomic sample, it will be ± 0.58 cycles, because $2^{-(0.58)} = 1.5$

To correct for slight differences in reaction efficiencies and detector dye intensities, the $\Delta\Delta C_T$ method, which relies on the analysis of a reference or calibrator sample, can be used *(21)*. Normally, the reference or calibrator sample is a sample of known (preferably normal) karyotype. The ΔC_T of the reference sample is subtracted from the ΔC_T of the sample, and then the corrected ratio of the target sequences in the sample can be calculated as follows:

$$\Delta\Delta C_{T\,\text{calibrated}} = \Delta C_T\,(\text{sample}) - \Delta C_T\,(\text{calibrator})$$

$$(\text{target A/target B})_{\text{calibrated}} = 2^{-(\Delta\Delta C_{T\,\text{calibrated}})}$$

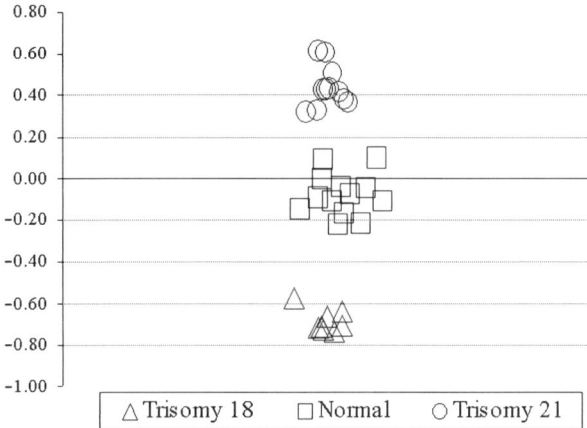

Fig. 3. Scatter plot of control DNAs for a threshold of 0.45. Depicted are replicate averages per sample. The change in threshold cycle (ΔC_T)-values of all samples with normal karyotype group between –0.22 and +0.10, all trisomy samples lie outside of this area (trisomy 21 are above, trisomy 18 are below). The average ΔC_T-value of the normal samples is –0.09. This value is used as the ΔC_T (calibrator) for samples with a normal karyotype at the threshold of 0.45 (**Table 1**).

Even though this method is more reliable than solely relying on the ΔC_T-value(s) of the analyzed sample, the use of a single reference sample can nevertheless introduce a small source of error. This may be caused by minute deviations in the amplification and signal generation, which may in turn lead to the misinterpretation of the sample analyzed. In this regard, it is worth noting that a seemingly small shift in the reference ΔC_T-value of only +0.15 cycle from its "real" value may result in misdiagnosis of a sample with a similarly small deviation of –0.15 cycle. For this reason, we recommend that for the reliable determination of ploidy an adaptation of the $\Delta\Delta C_T$ method be used, whereby the samples being analyzed are quantified relative to a reference ΔC_T-value determined from the average ΔC_T-values of all karyotypically normal samples tested previously. Furthermore, if this analysis suggests that a particular sample is aneuploid, then an added precaution may be to analyze this sample using a reference comprised of similar aneuploid samples. In this manner, an opposite pattern to that normally expected will emerge, in that the $\Delta\Delta C_T$ analysis using a normal ploidy reference should be close to the theoretical value of ±0.58 cycle, whereas the $\Delta\Delta C_T$ comparison using a matched aneuploid reference standard should be close to 0.

In reality, however, the experimentally measured differences in $\Delta\Delta C_T$ between normal and trisomic karyotypes is not 0.58, but ranges from 0.47 to 0.55 depending on threshold-value set and the type of trisomy. As a consequence, the calculated chromosome ratios will average 1.4 rather than 1.5 for a trisomic sample. Although the exact reason for this small anomaly is unclear, it is likely attributable to minor differences in amplification efficiencies or incomplete spectral separation of the dyes *(26)*.

Because the $\Delta\Delta C_T$-values themselves are indicative of the karyotype, i.e., close to 0 for normal and ±0.58 depending on the type of trisomy, it is not actually necessary to determine the ratio. This is particularly evident when taking an average $\Delta\Delta C_T$-value taken at the four independent threshold points (**Table 1**). Alternatively, it is possible to add these four independent $\Delta\Delta C_T$-values. In this case, a normal karyotype will have a value close to 0, whereas a trisomy 21 sample will be on the order of 2 and a trisomy 18 sample will have a value of approx –2. In this manner, it is possible to determine the karyotypes of the samples with a minimum of postrun work by setting the cut-off range around the average $\Delta\Delta C_T$ of a given karyotype. In our analyses, we therefore set a cut-off range of 0.25 cycle around the average measured $\Delta\Delta C_T$ of a given karyotype, because the measured $\Delta\Delta C_T$s between normal and trisomy are approx ±0.5. Accordingly, the $\Delta\Delta C_{T\ calibrated}$ of karyotypically normal samples will have a range of 0.00 ± 0.25. The accuracy of the test result can be confirmed by reanalyzing the suspected trisomic sample in comparison with a reference value made up of matched aneuploid samples. Furthermore, the test can be made more stringent by defining a smaller range.

An example of such a $\Delta\Delta C_T$ analysis, where trisomy 18 and 21 samples could be discriminated from one another as well as samples with normal ploidy for these chromosomes, is illustrated as a scatter plot in **Fig. 3**. This figure shows the clear segregation of the three different karyotypes. In addition, the manner in which the averaged ΔC_T-values of all relevant control samples can serve as calibrator or reference values is displayed in **Table 1**. In our hands, this simple procedure (and its modifications as described in under **Subheading 3.**) has been shown to be most reliable, producing consistent results over several experiments and months.

1.3. Differences Between the Use of the SDS 7700 and SDS 7000 Real-Time PCR "Taqman" Instruments

In our proof-of-principle study for the detection of trisomy 21 by real-time PCR, we used the Applied Biosystems Sequence Detection System 7700 (SDS 7700) instrument, which employs a complex laser array for the excitation of the fluorescent-labeled reporter molecules. In this study, in which we examined only a small number of samples ($n = 21$), we were able to correctly determine the karyotype in 10 of the 11 trisomy 21 samples analyzed. The analysis also included 10 dilutions of the trisomy 21 samples. In two of the control samples and in one sample with trisomy 21, no definitive assessment was pos-

Table 1
**The Calibrator Values for the Three Karyotypes
and the Four Thresholds Used**[a]

	ΔC_T calibrator values			
Threshold	0.2	0.3	0.45	0.675
Trisomy 18	−0.52	−0.55	−0.68	−0.73
Normal	0.01	−0.03	−0.09	−0.11
Trisomy 21	0.48	0.45	0.44	0.42

[a]The average ΔC_T-values of the control sample measurements
for each karyotype are the respective calibrator values.

sible as a result of the poor quality or inadequate quantity of the DNA sample analyzed.

In the interim period, Applied Biosystems has introduced the smaller, less costly and user-friendlier SDS 7000 to the market. This instrument obviates the need for a complex laser-based system, instead employing a halogen lamp for the excitation of fluorescent probes. After careful appraisal of this instrument, which we had determined to be as robust, reliable, and accurate as the more sophisticated and larger SDS 7700, we then adapted our assay in such a manner that it would permit the simultaneous analysis of trisomies 18 and 21. This was achieved by replacing the chromosome 12 (*glyceraldehyde-3-phosphate dehydrogenase* [*GAPDH*] gene) control amplicon of the original trisomy 21 assay with a sequence on chromosome 18. In order to increase the specificity of our new assay, we also examined the influence of various other parameters, such as PCR reaction mix composition and primer sequences and purity as well as different probe chemistries. These studies were undertaken to ensure that the amplification efficiencies of the two reactions (chromosome 18 and 21) were equal.

As described previously, reference calibration values for the two chromosomes being interrogated were determined by examining DNA samples with either normal, trisomy 18, or trisomy 21 karyotype (**Table 1**). Once we had established these reference calibration values, we performed a large-scale study of almost 100 clinical amniotic fluid samples, in which we examined the ploidy for these two chromosomes in a blinded manner (Zimmermann et al., manuscript in preparation). In more than 86% of the cases, the aneuploidy status for these two chromosomes was established correctly. In those instances where we were not successful, the analysis was either hindered by inadequate concentrations of template DNA or by the presence of inhibitors in the original DNA preparation (*see* **Note 7**).

These results clearly demonstrate the diagnostic potential of real-time PCR, whether for prenatal diagnosis or for other analyses involving gene duplications or loss, such as are frequently found in cancer. It is worth noting that although the quantitative fluorescent PCR analysis of STRs has been established and clinically implemented for the rapid prenatal analysis of the most common aneuploidies, the use of real-time PCR does have a few advantages in that it is more amenable to automation and, because it is a closed system, is less prone to contamination.

2. Materials
2.1. DNA Extraction

1. For the extraction of DNA from the amniotic fluid samples, Chelex 100 resin (biotechnology grade 100–200 mesh sodium form) from Bio-Rad (Reinach, Switzerland, cat. no. 142-1253) was used.
2. It is recommended that a highly pure water source, such as Analar water (molecular biology grade, VWR International), be used. Alternatively, Milli-Q water deionized with the Elgastat Maxima (Kleiner AG, Wohlen, Switzerland) can be used.

2.2. Real-Time PCR

1. For our real-time PCR analysis, we used an ABI PRISM 7000 Sequence Detection System (SDS7000) (*see* **Note 1**).
2. For the PCR reactions, we used MicroAmp Optical 96-well reaction plates (Applied Biosystems, Switzerland, cat. no. 4306737) and the ABI PRISM Optical adhesive covers (cat. no. 4311971) (*see* **Note 2**).
3. The reactions were carried using the TaqMan Universal PCR master mix (cat. no. 4304437) containing Ampli*Taq* Gold DNA Polymerase, AmpErase uracil-*N*-glycosylase (UNG), dNTPs with dUTP, and passive Reference 1 (ROX fluorescent dye) (*see* **Note 3**).
4. Primer and probe sequences are listed in **Table 2** (*see* **Note 4**).
5. The best results were obtained using TaqMan minor groove-binding (MGB) probes (cat. no. 43160324) from ABI. These are 3'-labeled with a minor groove binder and a nonfluorescent quencher (*see* **Note 5**). The chromosome 18 probe is 5'-labeled with the fluorescent dye VIC, and the chromosome 21 probe is 5'-labeled with FAM. It is advisable to store these probes as 5-μM aliquots at $-20°C$, at which they are stable for over 1 yr. If used rapidly (within 1 mo), the probes can be stored at 4°C. In this case, aliquots of both probes should be used and stored in parallel. Avoid exposure to light.
6. In our experience, the best results were obtained using high-performance liquid chromoatography (HPLC)-purified primers (Microsynth, Switzerland) (*see* **Note 6**). These can be stored as 10- μM aliquots at $-20°C$. Primers stored at 7°C are stable for several months.

Table 2
Primer and Probe Sequences for the Real-Time Amplification of Chromosomes 18 and 21[a]

Chromosome location	Accession GenBank	Gene	Name	Fluorescent dyes 5'–3'	Sequence of primers	Primer location bases
18p11.32	D00596	TYMS	Chr_18_F		TGACAACCAAACGTGTGTTCTG	2910–2931
			Chr_18_R		AGCAGCGACTTCTTTACCTTGATAA	2961–2985
			Chr_18_P_MGB	VIC-NFQ	GGTGTTTGGAGGAGTT	2933–2959
			Chr_18_P_dual	VIC-TAMRA	AAGGGTGTTTGGAGGAGTTGCTGTGG	2930–2966
21q21.3	D87675	APP	Chr_21_F		CCCAGGAAGGAAGTCTGTACCC	7764–7785
			Chr_21_R		CCCTTGCTCATTGCGCTG	7826–7843
			Chr_21_P_MGB	FAM-NFQ	CTGGCTGAGCCATC	7788–7801
			Chr_21_P_dual	FAM-TAMRA	AGCCATCCTTCCCGGGCCTAGG	7795–7816

[a]Probes used were the minor groove-binding (MGB) probes Chr_18_P_MGB and Chr_21_P_MGB. The sequences for dual labeled probes are included (*see* **Note 5**).

3. Methods

3.1. DNA Extraction of Amniotic Fluid Samples

1. Spin 1 mL of amniotic fluid at 16,000*g* in a benchtop micro-centrifuge for 15 s to pellet the amniotic cells.
2. Discard the supernatant and resuspend the cell pellet in 500 µL of water.
3. Wash once by centrifuging the cells as in **step 1**. Discard the supernatant.
4. Add 60 µL of Chelex 100 resin to the pellet and resuspend the cells by vortexing.
5. Incubate the samples for 20 min at 57°C.
6. Incubate the samples in a dry heat block at 100°C for 8 min.
7. Prepare the sample for use in the real-time PCR analysis by heating to 95°C for 5 min. Following this, spin the samples at 16,000*g* in a benchtop micro-centrifuge for 2 min to pellet the resin. Carefully remove the aqueous DNA solution, taking care not to disturb the resin pellet (*see* **Note 7**).
8. Use 2 µL of this DNA solution in a final reaction volume of 25 µL for the PCR analysis (*see* **Notes 8** and **9**).

3.2. Performing Real-Time PCR

1. It is advisable to include at least one control sample of known karyotype in the analytic run (*see* **Note 10**).
2. Each sample should be analyzed at least in triplicate (*see* **Note 11**).
3. The preparation of the real-time PCR reactions should be carried out on ice.
4. The real-time PCR amplification is carried out in a total volume of 25 µL (*see* **Note 12**) containing 2 µL of the sample DNA solution (*see* **Note 13**), 300 n*M* of each primer (*see* **Note 14**), and 200 n*M* of each probe (*see* **Note 15**) at 1X concentration of the Universal PCR reaction mix.
5. This is prepared by first pipetting the PCR reaction mixture into the reaction wells. Follow this step by adding 2 µL of the sample DNA solution (*see* **Note 16**). Carefully seal the reaction plate with the optical adhesive cover. Centrifuge at 1000*g* at 4°C for 1 min to spin down any droplets and remove air bubbles.
6. Immediately start the real-time PCR cycler with the emulation mode off (*see* **Note 17**).
7. Following an initial incubation at 50°C for 2 min to permit Amp Erase activity, and another at 95°C for 10 min to allow activation of Ampli*Taq* Gold and denaturation of the genomic DNA, use the following cycle conditions: 40 cycles of 1 min at 60°C and 15 s at 95°C.

3.3. Analysis of Real-Time PCR Data

1. Perform the experimental analysis with a baseline setting of 3–22 cycles (*see* **Note 18**). Check the replicate curves for uniformity in the amplification plot view (*see* **Note 19**) and for final fluorescence (*see* **Note 20**).
2. Analyze four different thresholds at R_n-values of: 0.2, 0.3, 0.45, and 0.675, as indicated in **Fig. 2** and **Table 1** (*see* **Note 21**). Export the C_T result files and examine the data using an Excel spreadsheet.

3. In our experience, C_T-values from 25.0 to 29.0 (for a threshold of 0.2) will yield good results. C_T-values of higher than 29.0 are associated with quantities of DNA that are too low to guarantee that preferential amplification of one of the multiplexed reactions does not take place (*see* **Note 22**).
4. For every replicate, calculate the ΔC_T for each of the four thresholds (**Table 3** and **Fig. 2**).
5. For every replicate and for all three ploidies that can be distinguished, calculate the $\Delta\Delta C_T$ for each of the four thresholds:

$$\Delta\Delta C_T \text{ calibrated} = \Delta C_T \text{ (sample)} - \Delta C_T \text{ (calibrator)}.$$

6. Calculate for every replicate, with respect to a normal karyotype and the two types of trisomies, the average of the four $\Delta\Delta C_T$-values.
7. The analysis of one replicate that gives the value closest to 0 for the averaged $\Delta\Delta C_T$ is indicative of the chromosomal status. This value should be smaller than ± 0.25 (*see* **Note 23**); that is, for example:

$$\Delta\Delta C_T \text{ calibrated (trisomy 18 sample)} = \Delta C_T - \Delta C_T \text{ (calibrator trisomy 18)} = 0 \pm 0.25.$$

8. If the results of all three replicates are indicative of the same chromosomal balance, the ploidy of the sample is that of the matched reference ΔC_T calibrators; otherwise, the result is not conclusive but rather is suggestive (*see* **Note 24**).
9. Alternatively, it is possible to add the four calibrated $\Delta\Delta C_T$-values. In this case, a normal karyotype will have a value close to 0, whereas a trisomy 21 sample will be on the order of 2 and a trisomy 18 sample will have a value of approx -2.

4. Notes

1. In order for this assay to work, the main requirement for the real-time PCR instrument is that the fluorescence data are normalized to the passive reference dye, such as the ROX dye used by ABI. Any instrument that offers this feature should be suited for the test. In a previous study, we used the ABI PRISM 7700 Sequence Detection System successfully.
2. The optical covers have a high light-transmission capacity, a necessity for highly precise measurements. Although optical caps should also work (according to the manufacturers' technical support), our experience shows that the automatic data collection adjustments of the SDS7000 are not sufficient for achieving the accuracy needed. For the laser-illuminated SDS7700, the covers and caps work fine. However, it is advisable to always use the same covering for the reaction wells, because it has an influence on the measured fluorescence.
3. The use of AmpErase UNG is absolutely required in order to avoid amplification of any misprimed, nonspecific products formed prior to specific amplification. If this step is omitted, the balance of the two multiplexed reactions can be lost and one of the sequences will be amplified with a greater efficiency, resulting in false results. For the same reason, use of a *Taq* DNA polymerase activated by Hot Start is required.
4. To match the two assays, a variety of primer combinations for the chromosome 21 were tested. For the sake of expedience, we have listed only the most efficient ones as well as the pertinent MGB probes (**Table 2**).

Table 3
Experimental Data and Analysis of Three Samples[a]

	Sample 1				Sample 2				Sample 3			
Threshold	0.2	0.3	0.45	0.675	0.2	0.3	0.45	0.675	0.2	0.3	0.45	0.675
C_T (chromosome 18)	27.83	28.47	29.25	30.17	27.91	28.47	29.17	29.96	27.21	27.81	28.45	29.28
C_T (chromosome 21)	28.07	28.66	29.34	30.15	28.53	29.24	30.00	31.01	26.65	27.28	28.01	28.80
ΔC_T	−0.24	−0.19	−0.09	0.02	−0.62	−0.77	−0.83	−1.05	0.56	0.53	0.44	0.48
$\Delta\Delta C_T$ (Normal)	−0.20	−0.10	0.06	0.25	−0.58	−0.68	−0.68	−0.82	0.60	0.62	0.59	0.71
$\Delta\Delta C_T$ (Trisomy 18)	0.30	0.40	0.61	0.79	−0.08	−0.18	−0.13	−0.28	1.10	1.12	1.14	1.25
$\Delta\Delta C_T$ (Trisomy 21)	−0.68	−0.57	−0.45	−0.30	−1.06	−1.15	−1.19	−1.37	0.12	0.15	0.08	0.16
Avg $\Delta\Delta C_T$ (Norm)	**0.00**				−0.69				0.63			
Avg $\Delta\Delta C_T$ (T 18)	0.53				**−0.17**				1.15			
Avg $\Delta\Delta C_T$ (T 21)	−0.50				−1.19				**0.13**			

[a]Single replicates. The average $\Delta\Delta C_T$-values of the determined karyotypes are in bold. From this table, it is apparent that sample 1 has a normal karyotype, sample 2 is of trisomy 18, and sample 3 is trisomy 21 (it is one of the replicates depicted in **Figs. 1–3**).

5. MGB probes emit lower background fluorescence than dual-labeled probes. This is mainly owing to more efficient quenching because the probes are shorter. As a result, MGB probes have a longer observable exponential phase, and this permits the use of a wider range for the thresholds. Dual-labeled probes may also be used, but the measurements are slightly inferior in terms of final accuracy.

   It is also important to be aware that the use of different batches of the same probes can result in differences in measurement because of minor differences in quality.

6. It is highly recommended that HPLC purified primers be used, as these are not subject to artifacts that nonpurified primers are prone to, such as the formation of unspecific products by the presence of shorter unspecific primer fragments, which have an adverse effect on the reaction efficiency. Nonpurified primers are also subject to greater batch-to-batch variation than HPLC purified ones.

7. It is very important that the Chelex resin suspension be heated prior to centrifugation and separation of the DNA from the resin pellet. If not, readsorption of the DNA by the resin can occur, thereby removing a major amount of the DNA from the solution. This adsorption process may also lead to an imbalance of the ratio of the two target chromosomes. It is also important that no resin is transferred to the PCR as this will adversely affect amplification.

8. DNA concentrations need not be quantified. The amount of DNA needed for highly reproducible results ranges from 5 to 80 ng. The protocol will usually produce a DNA solution in the required concentration range. Because of the limited quantity of sample material available, it should not be used for concentration measurements. The additional working step is not necessary.

   By co-amplifying a reference sample of known amount of DNA, which is the usual practice, the concentration of the sample can be determined with the following equation:

$$\text{concentration of sample} = 2^{-(\Delta CT)}/\text{concentration reference sample.}$$

9. Any kind of genomic DNA where the chromosomal ratio is preserved can be used. In addition to the Chelex-extracted amniotic fluids, we also tested extractions with the highly pure PCR template kit from Roche and with the QIAmp blood kits from Qiagen. Cellular materials used included whole blood, monocytes from blood, and cultured cells from blood and from amniotic fluid.

10. Use of a reference sample of known quantity allows quantification, as mentioned in **Note 8**. If the sample and a dilution of it are amplified, the sequence detection software can determine the concentration of all samples tested with two standard curves (one for each chromosome).

    In theory, these curves would permit karyotyping of the samples. In practice, however, the results are not good enough for karyotyping.

11. It is a good idea to amplify both the undiluted and a diluted sample. PCR inhibitors may sometimes be present in the original sample. Such inhibitors will then be diluted during the diluting process. This added precaution makes it possible to observe reaction efficiencies for a sample via a standard curve and, thus, reveal the presence of inhibitors.

12. Reaction volumes between 20 and 50 µL had been tested. The system will work with 20 µL, but the deviations are increased. A volume of 50 µL works well, but it is not always necessary to use such a large volume. Very small differences in collected data might occur if the reaction volume is different from the volume that the assay is characterized for.

13. If the concentration of the DNA is too low, up to 4 µL Chelex-extracted solution can be used. If the concentration is still too low, the DNA solution must be concentrated by, for example, evaporation, precipitation, or extraction into a smaller volume using a kit.

    To save DNA and for highly concentrated samples, smaller volumes or dilutions in water can be used.

14. Higher primer concentrations result in a decrease of the reaction efficiency in the multiplex assay. This can be attributed to an increase of unspecific reactions. In addition, excessively high concentrations of genomic DNA are detrimental to the reaction.

15. Probe concentrations of 100 µM also work fine, but a little accuracy may be lost, especially at the threshold of 0.675.

16. Sample setup on ice is important in order to minimize nonspecific product formation. The reaction plate should be kept on ice at all times prior to starting the real-time PCR.

17. Run the program always at the same heating and cooling rates. Differences in the temperature program will result in slight differences of the final C_T- and ΔC_T-values. We use the "emulation off" setting because it is faster.

18. The baseline setting is very important in order to achieve optimal results in the analysis. It is used to separate unspecific fluorescent signal from the signal generated by the reaction. If the baseline is too far away from the amplification curve, a significant proportion of the results, especially those with lower thresholds, will be adversely affected by noise.

    Setting the baseline too close will skew the amplification curves and change the ΔC_T to smaller values.

19. In the amplification plot view, check whether the amplification curves of both dyes look normal and parallel for all three replicates. If the three replicates do not look very similar, something may be wrong with at least one of the reactions. This can be owing to contamination with DNA (pipetting errors), dirt or fluorescence source other than the dyes, leakage of the well, presence of an inhibitor or other reasons. If the curves do not resemble each other closely, the test should be re-run.

20. After 40 cycles of amplification, the ΔR_n will usually be between 2.0 and 3.0. Runs with a final ΔR_n of less than 1.5 are generally of poor quality.

    If using dual-labeled probes, the ΔR_n will be smaller than when using MGB probes. This is a result of less efficient quenching and resulting higher background fluorescence in the longer dual-labeled probes.

21. The use of several thresholds increases the accuracy of the method. By using the four proposed thresholds, fluorescence data can be collected during three or four cycles; compare with the use of one threshold, in which case data is collected from only two cycles. A threshold of 0.1 sometimes contains a good amount of

unspecific signal. To exclude that possibility, the minimal threshold setting used is 0.2. The maximum threshold of 0.675 still gives good data, although the reaction is at the end of the exponential phase and efficiencies are decreasing.

22. At DNA concentrations that result in a C_T of 30.0 or higher (at the threshold of 0.2), deviations occur in the amplifications, sometimes resulting in ΔC_T-values uncharacteristic for the sample karyotype. Using a cutoff of 29.0 is a security measure to guarantee balanced amplification of both targets.

   The cutoff of 25.0 is a result of the baseline setting. Also, the amplification of genomic DNA of very high concentrations can result in lower and varying efficiencies. None of the 100 amniotic fluid samples tested had a high enough concentration to result in such a low C_T.

23. To make the test more stringent, the maximum average $\Delta\Delta C_T$-value can be set to a smaller value. An alternative stringency requirement could be to demand that the results of a minimal number of thresholds per sample are indicative of the normal karyotype or one of the trisomies (e.g., 10 of total 12 values per triplicate).

24. It is usually possible to exclude one of the three karyotypes.

References

1. Dudarewicz, L., Holzgreve, W., Jeziorowska, A., Jakubowski, L., and Zimmermann, B. (2005) Molecular methods for rapid detection of aneuploidy. *J. Appl. Genet.* **46,** 207–215.
2. Klinger, K., Landes, G., Shook, D., et al. (1992) Rapid detection of chromosome aneuploidies in uncultured amniocytes by using fluorescence in situ hybridization (FISH). *Am. J. Hum. Genet.* **51,** 55–65.
3. Witters, I., Devriendt, K., Legius, E., et al. (2002) Rapid prenatal diagnosis of trisomy 21 in 5049 consecutive uncultured amniotic fluid samples by fluorescence in situ hybridisation (FISH). *Prenat. Diagn.* **22,** 29–33.
4. Philip, J., Bryndorf, T., and Christensen, B. (1994) Prenatal aneuploidy detection in interphase cells by fluorescence in situ hybridization (FISH). *Prenat. Diagn.* **14,** 1203–1215.
5. Mutter, G. L. and Pomponio, R. J. (1991) Molecular diagnosis of sex chromosome aneuploidy using quantitative PCR. *Nucleic Acids Res.* **19,** 4203–4207.
6. von Eggeling, F., Freytag, M., Fahsold, R., Horsthemke, B., and Claussen, U. (1993) Rapid detection of trisomy 21 by quantitative PCR. *Hum. Genet.* **91,** 567–570.
7. Levett, L. J., Liddle, S., and Meredith, R. (2001) A large-scale evaluation of amnio-PCR for the rapid prenatal diagnosis of fetal trisomy. *Ultrasound Obstet. Gynecol.* **17,** 115–118.
8. Mansfield, E. S. (1993) Diagnosis of Down syndrome and other aneuploidies using quantitative polymerase chain reaction and small tandem repeat polymorphisms. *Hum. Mol. Genet.* **2,** 43–50.
9. Bernard, P. S. and Wittwer, C. T. (2002) Real-time PCR technology for cancer diagnostics. *Clin. Chem.* **48,** 1178–1185.
10. Bustin, S. A. (2002) Quantification of mRNA using real-time reverse transcription PCR (RT-PCR): trends and problems. *J. Mol. Endocrinol.* **29,** 23–39.

11. Dotsch, J., Repp, R., Rascher, W., and Christiansen, H. (2001) Diagnostic and scientific applications of TaqMan real-time PCR in neuroblastomas. *Expert Rev. Mol. Diagn.* **1,** 233–238.

12. Ginzinger, D. G. (2002) Gene quantification using real-time quantitative PCR: an emerging technology hits the mainstream. *Exp. Hematol.* **30,** 503–512.

13. Muller, P. Y., Janovjak, H., Miserez, A. R., and Dobbie, Z. (2002) Processing of gene expression data generated by quantitative real-time RT-PCR. *Biotechniques* **32,** 1372–1374, 1376, 1378–1379.

14. Hahn, S., Zhong, X. Y., Burk, M. R., Troeger, C., and Holzgreve, W. (2000) Multiplex and real-time quantitative PCR on fetal DNA in maternal plasma. A comparison with fetal cells isolated from maternal blood. *Ann. N. Y. Acad. Sci.* **906,** 148–152.

15. Zimmermann, B., El-Sheikhah, A., Kypros, N., Holzgreve, W., and Hahn, S. (2005) Optimized real-time quantitative PCR measurement of male fetal DNA in maternal plasma. *Clin. Chem.* **51,** 1598–1604.

16. Zimmermann, B., Holzgreve, W., Zhong, X. Y., and Hahn, S. (2002) Inability to clonally expand fetal progenitors from maternal blood. *Fetal Diagn. Ther.* **17,** 97–100.

17. Zhong, X. Y., Holzgreve, W., and Hahn, S. (2000) Detection of fetal Rhesus D and sex using fetal DNA from maternal plasma by multiplex polymerase chain reaction. *BJOG* **107,** 766–769.

18. Zimmermann, B., Holzgreve, W., Wenzel, F., and Hahn, S. (2002) Novel real-time quantitative PCR test for trisomy 21. *Clin. Chem.* **48,** 362–363.

19. Meijerink, J., Mandigers, C., van de Locht, L., Tonnissen, E., Goodsaid, F., and Raemaekers, J. (2001) A novel method to compensate for different amplification efficiencies between patient DNA samples in quantitative real-time PCR. *J. Mol. Diagn.* **3,** 55–61.

20. Livak, K. J. (1997) Relative quantification of gene expression. *ABI Prism 7700 Sequence Detection System user bulletin 2.* PE Applied Biosystems, Foster City, CA.

21. Livak, K. J. and Schmittgen, T. D. (2001) Analysis of relative gene expression data using real-time quantitative PCR and the $2^{(-\Delta\Delta C_T)}$ Method. *Methods* **25,** 402–408.

22. Pfaffl, M. W. (2001) A new mathematical model for relative quantification in real-time RT-PCR. *Nucleic Acids Res.* **29,** e45.

23. Ramakers, C., Ruijter, J. M., Deprez, R. H., and Moorman, A. F. (2003) Assumption-free analysis of quantitative real-time polymerase chain reaction (PCR) data. *Neurosci. Lett.* **339,** 62–66.

24. Peccoud, J. and Jacob, C. (1998) Statistical estimations of PCR amplification rates, in *Gene Quantification* (Ferre, F., ed.). Birkhauser, Boston: pp. 111–128.

25. Liu, W. and Saint, D. A. (2002) Validation of a quantitative method for real time PCR kinetics. *Biochem. Biophys. Res. Commun.* **294,** 347–353.

26. Mathieu-Daude, F., Welsh, J., Vogt, T., and McClelland, M. (1996) DNA rehybridization during PCR: the 'Cot effect' and its consequences. *Nucleic Acids Res.* **24,** 2080–2086.

9

Noninvasive Prenatal Diagnosis by Analysis of Fetal DNA in Maternal Plasma

Rossa W. K. Chiu and Y. M. Dennis Lo

Summary

Prenatal diagnosis has become a standard part of obstetrics care. Genetic diagnoses are established prenatally through the sampling of fetal genetic material by invasive methods such as amniocentesis or chorionic villus sampling, which are associated with a risk of fetal loss. Hence, the recent discovery of the presence of fetal DNA in maternal plasma has led to exciting possibilities of noninvasive prenatal diagnosis. Numerous applications have since been reported for the analysis of circulating fetal DNA. The accuracy of fetal DNA assessment from maternal plasma is dependent on the appropriate preanalytical handling of maternal blood samples, an efficient fetal DNA extraction protocol, and a sensitive and specific detection method. The protocol that has been applied regularly in the laboratory of the authors for the reliable detection and quantification of circulating fetal DNA is presented in this chapter.

Key Words: Fetal DNA; prenatal diagnosis; noninvasive; real-time quantitative PCR; circulating nucleic acids.

1. Introduction

Prenatal diagnosis is an indispensable part of present-day obstetrics care. Prenatal genetic diagnosis relies on the sampling of fetal tissues and at present is dependent on procedures such as amniocentesis or chorionic villus sampling. As these procedures are invasive in nature, they are associated with a small but finite risk of spontaneous abortion. In 1997, the existence of fetal DNA in maternal plasma was first reported and has opened up exciting opportunities for the development of noninvasive methods for prenatal diagnosis (1). It has subsequently been demonstrated that significant amounts of fetal DNA are present in maternal plasma, amounting to 3–6% of the total DNA in maternal plasma (2). With the adoption of sensitive molecular methods, such as real-

From: *Methods in Molecular Biology, vol. 336: Clinical Applications of PCR*
Edited by: Y. M. D. Lo, R. W. K. Chiu, and K. C. A. Chan © Humana Press Inc., Totowa, NJ

time quantitative polymerase chain reaction (PCR), fetal DNA can be reliably detected in maternal plasma between the fifth to seventh week of gestation *(2–4)*. Circulating fetal DNA concentration increases with progression of pregnancy and disappears from maternal plasma rapidly after delivery, with a median half-life of 16.3 min *(2,5–7)*. Hence, circulating fetal DNA represents a promising source of fetal genetic material for prenatal diagnosis while being accessible simply by maternal venesection.

Soon after its discovery, numerous potential applications have been reported for the analysis of fetal DNA in maternal plasma. Abnormal concentrations of circulating fetal DNA had been reported for preeclampsia *(8,9)*, fetal aneuploidy *(10,11)*, preterm labor *(12)*, fetal hemorrhage *(13)*, invasive placentation *(14,15)*, hyperemesis gravidarum *(16)*, and oligohydramnios *(17)*. Mutation detection through circulating fetal DNA analysis had been reported for a number of hereditary traits including fetal rhesus D status *(18,19)*, β-thalassemia *(20,21)*, congenital adrenal hyperplasia *(22)*, and achondroplasia *(23)*.

Because of the relative high abundance, fetal DNA can be readily detected in maternal plasma with PCR amplification, particularly when a number of preanalytical and analytical considerations are taken into account *(24)*. These considerations include processes involving the choice of specimen, timing and procedures for blood processing, design and format of assay system, and anti-contamination procedures.

2. Materials

2.1. Sample Collection

Ethylenediamine tetraacetic acid (EDTA)-containing blood collection tubes for plasma collection.

2.2. DNA Extraction

1. Absolute ethanol.
2. QIAamp Blood Mini Kit (Qiagen, Hilden, Germany).
   a. Protease.
   b. Lysis buffer.
   c. Spin columns.
   d. Collection tubes.
   e. Wash buffer 1.
   f. Wash buffer 2.
3. Deionized water (dH$_2$O).

2.3. Real-Time Quantitative PCR

2.3.1. Amplification Reagents (2)

1. Primers (*see* **Note 1**):

    a. *SRY*:
       SRY-109F: 5'-TGG CGA TTA AGT CAA ATT CGC-3'.
       SRY-245R: 5'-CCC CCT AGT ACC CTG ACA ATG TAT T-3'.
    b. β-*globin*:
       β-globin-354F: 5'-GTG CAC CTG ACT CCT GAG GAG A-3'.
       β-globin-455R: 5'-CCT TGA TAC CAA CCT GCC CAG-3'.
2. Dual-labeled fluorescent probes (*see* **Note 1**):
    a. *SRY*:
       SRY-142T: 5'-(FAM)AGC AGT AGA GCA GTC AGG GAG GCA GA(TAMRA)-3'.
    b. β-*globin*:
       β-globin-402T: 5'-(FAM)AAG GTG AAC GTG GAT GAA GTT GGT GG(TAMRA)-3'.
    where FAM is 6-carboxy-fluorescein; TAMRA is 6-carboxy-tetramethyl-rhodamine.
3. Calibrator: male genomic DNA.
4. TaqMan PCR core reagent kit (Applied Biosystems, Foster City, CA).
    a. 10X buffer A.
    b. MgCl$_2$.
    c. dATP, dCTP, dGTP, dUTP.
    d. Ampli*Taq* Gold.
    e. AmpErase uracil *N*-glycosylase.
5. dH$_2$O.

2.3.2. Instrumentation for Quantitative Analysis

ABI Prism 7700 Sequence Detector (Applied Biosystems).

3. Methods

The methods described below outline the procedures for (1) sample collection, (2) sample processing, (3) maternal plasma DNA extraction, (4) construction of calibration curves, (5) real-time quantitative PCR analysis, (6) data interpretation, and (7) prevention of contamination.

3.1. Sample Collection

Collect at least 2 mL of maternal venous blood into EDTA tubes (*see* **Note 2**).

3.2. Sample Processing

1. Maternal blood samples may be kept at room temperature or 4°C until processing (*see* **Note 3**).
2. Process maternal whole blood within 6 h to obtain plasma (*see* **Note 3**).
3. Centrifuge maternal whole blood at 1600*g* for 10 min.
4. Transfer supernatant to clean polypropylene tubes with due caution not to disturb the underlying buffy coat layer.

5. Centrifuge supernatant at 16,000g for 10 min *(25)* (*see* **Note 4**).
6. Transfer plasma (supernatant) to clean polypropylene tubes with caution not to disturb the cell pellet.
7. Store plasma at –20°C until DNA extraction.

3.3. Maternal Plasma DNA Extraction

Follow the "blood and body fluid protocol" of the QIAamp Blood Mini kit (Qiagen) with the following modifications:

1. 800 µL of maternal plasma is taken from each sample for DNA extraction. The maternal plasma from each sample is first divided into two 400-µL portions and transferred into separate polypropylene tubes. Both aliquots are processed.
2. The "blood and body fluid protocol" is followed but scaled to process 400 µL of samples, i.e., 40 µL of protease and 400 µL of lysis buffer added to each 400- µL aliquot of maternal plasma.
3. Incubate the solution at 56°C for 10 min.
4. Add 500 µL of cold absolute ethanol to each aliquot.
5. Transfer 630 µL of the ethanol mixture from one of the paired aliquots to a spin column with care not to contaminate the cap or surrounding areas of the spin column.
6. Centrifuge at 16,000g for 1 min.
7. Discard infranatant, place the spin column into a new collection tube.
8. Repeat **steps 5–7** three times, until all of the ethanol mixture of the two paired aliquots have been applied to the same spin column.
9. Add 500 µL of reconstituted Wash Buffer 1 to each spin column and centrifuge at 16,000g for 1 min.
10. Discard infranatant and place the spin column in a new collection tube. Add 500 µL of reconstituted Wash Buffer 2 to each spin column and centrifuge at 16,000g for 3 min.
11. Discard infranatant. Transfer spin column to a clean polypropylene tube.
12. Add 50 µL of dH$_2$O to each spin column.
13. Incubate at room temperature for 5 min.
14. Centrifuge at 16,000g for 1 min. Collect the extracted DNA (infranatant) and store at –20°C until analysis. Discard spin columns.

3.4. Construction of Calibration Curves

1. Determine the DNA content of a male genomic DNA extract by optical density measurement (*see* **Note 5**). Calculate the DNA content in terms of genomic equivalence of the stock calibrator using a conversion factor of 6.6 pg of DNA per cell.
2. Serially dilute the stock calibrator to prepare a set of calibrators consisting of standards with 0.78, 1.56, 3.12, 6.25, 12.5, 25, 100, 500, 1000, 2500, and 10,000 genome equivalents (GE)/5 µL.
3. Store the standards at –20°C.

3.5. Real-Time Quantitative PCR Analysis (see Note 6)

1. Prepare reagent master mix according to manufacturer's instructions for the TaqMan PCR Core reagent kit (Applied Biosystems). The reagent mix consists of 5 μL 10X buffer A; 300 n*M* of each amplification primer, 100 n*M* of the corresponding fluorescent probe, 4 m*M* MgCl$_2$, 200 μ*M* each dATP, dCTP, dGTP; 400 μ*M* dUTP; 1.25 U Ampli*Taq* Gold; and 0.5 U AmpErase uracil *N*-glycosylase (2). Add dH$_2$O to make up to a volume of 45 μL per reaction. Thoroughly mix the solution by vortexing.
2. Dispense 45 μL of reagent mix to each reaction well on a 96-well optical plate. Perform duplicate analysis for each sample or standard (*see* **Note 7**). Allow at least two to four reactions as no-template control (NTC). Add known male and female DNA samples as positive and negative controls.
3. Add 5 μL of sample, standard or dH$_2$O (for NTC), to the corresponding reaction well.
4. Take care to remove air bubbles from the sample-reagent mix before sealing the optical plate. Ensure that the optical plate is sealed securely.
5. Real-time quantitative PCR is carried out in an Applied Biosystems 7700 Sequence Detector. Set the thermal profile as follows: 2 min incubation at 50°C, followed by a first denaturation of 10 min at 95°C, then 40 cycles of 95°C for 15 s and 60°C for 1 min (2).

3.6. Data Interpretation

1. Confirm that the NTC and negative controls are negative. Confirm that the positive controls are positive. Assess the linearity of the standard curve by confirming whether the correlation coefficient is better than 0.95. Confirm that the run has a detection sensitivity of one to three copies. Proceed to further interpretation of the PCR analysis when the above conditions are fulfilled.
2. The number of copies of DNA template (*Q*) at the start of the reaction will be computed by the sequence detection software (Applied Biosystems). Determine the fetal (*SRY*) or maternal (β-*globin*) DNA concentrations (*C*) in the original maternal plasma samples by the following calculation:

$$C = Q \times \frac{V_{DNA}}{V_{PCR}} \times \frac{1}{V_{ext}} \tag{2}$$

where *C* = DNA concentration in plasma (GE/mL); *Q* = plasma DNA quantity (copies) determined by the sequence detector in a PCR; V_{DNA} = total volume of plasma DNA obtained after extraction, typically 50 μL per extraction; V_{PCR} = volume of plasma DNA solution used for PCR, typically 5 μL; and V_{ext} = volume of plasma extracted, typically 800 μL.

3.7. Anti-Contamination Procedures

1. Perform the sample processing, DNA extraction, PCR preparation, and real-time PCR in separate designated work areas. Prepare the PCR reagent mix within a laminar flow cabinet.
2. Use aerosol-resistant pipet tips.
3. Incorporate the use of uracil *N*-glycosylase to each PCR to eliminate potentially contaminating amplicons from previous PCR reactions.

4. Notes

1. *SRY* is located on the Y-chromosome and therefore is a fetal-specific target for plasma of women pregnant with male fetuses. β-*globin* is present in both the fetal and maternal genomes, and, therefore, its concentration reflects the total concentration of DNA in maternal plasma. The *SRY* and β-*globin* real-time PCR assays are described as illustrative examples for the detection of fetal and total DNA in maternal plasma, respectively. Other fetal targets detectable in maternal plasma include *RHD (18)* and fetal-specific mutant alleles *(20)*.
2. Fetal DNA has been detectable in both maternal plasma and serum *(1,2,26)*. However, the background maternal DNA in maternal serum is 14.6 times higher than that in paired plasma samples from corresponding subjects *(2)*. The excess serum DNA is contributed by blood clotting *(27,28)*. The excess maternal background DNA translates to a 26- and 6-fold decrease in maternal serum fetal DNA concentration compared with maternal plasma in first and third trimesters, respectively *(2)*. We have further observed (unpublished) that the analysis of fetal-specific mutations revealed much higher false-negative detection in maternal serum than plasma, possibly a consequence of the reduced fractional concentration of fetal DNA in maternal serum.
3. Studies have shown that prolonged storage of unprocessed whole blood leads to lysis of blood cells with resultant elevations in maternal DNA concentrations *(28,29)*. The effect is much more apparent in plain whole blood than anticoagulated blood. Significant elevation in serum DNA concentration can be demonstrated with a 2-h delay of blood processing *(29)*. However, concentration of maternal DNA derived from plasma with EDTA as anticoagulant remains stable for 6 h *(30)*. These data further support the advantage of maternal plasma over serum for fetal DNA analysis and also suggest that maternal blood samples collected into EDTA tubes should be processed within 6 h.
4. Previous studies have revealed that the way plasma is derived from maternal whole blood has important implications on the analysis of maternal plasma DNA *(24)*. It has been shown that to ensure consistent and reliable analytical results, cellular elements should be eliminated from maternal plasma either by adopting a two-step centrifugation protocol or filtration of plasma *(25,31)*. Contamination of plasma with residual cellular elements could result in a fluctuant background of maternal DNA *(25)*, affect the quantitative interpretation of maternal plasma total DNA concentration *(25)*, and cause the inadvertent detection of fetal cells that have persisted in maternal plasma from previous pregnancies *(32,33)*.

5. Alternative calibrant materials, such as synthetic oligonucleotides *(34)* or plasmid with inserts of targeted genes *(35)*, have also been reported.

6. The protocol focuses on the use of real-time quantitative PCR for fetal DNA detection in maternal plasma. However, other assay platforms have also been adopted for fetal DNA detection, including conventional PCR, which is less sensitive *(3)*; and mass spectrometry, which has both superior sensitivity and specificity *(36)*.

7. Many groups perform triplicate analyses *(30,37)*, and it has been reported that analytical results can be interpreted with greater confidence when higher number of replicates are performed *(38)*.

References

1. Lo, Y. M. D., Corbetta, N., Chamberlain, P. F., et al. (1997) Presence of fetal DNA in maternal plasma and serum. *Lancet* **350,** 485–487.

2. Lo, Y. M. D., Tein, M. S., Lau, T. K., et al. (1998) Quantitative analysis of fetal DNA in maternal plasma and serum: implications for noninvasive prenatal diagnosis. *Am. J. Hum. Genet.* **62,** 768–775.

3. Birch, L., English, C. A., O'Donoghue, K., Barigye, O., Fisk, N. M., and Keer, J. T. (2005) Accurate and robust quantification of circulating fetal and total DNA in maternal plasma from 5 to 41 weeks of gestation. *Clin. Chem.* **51,** 312–320.

4. Gonzalez-Gonzalez, C., Garcia-Hoyos, M., Trujillo-Tiebas, M. J., et al. (2005) Application of fetal DNA detection in maternal plasma: a prenatal diagnosis unit experience. *J. Histochem. Cytochem.* **53,** 307–314.

5. Lo, Y. M. D., Zhang, J., Leung, T. N., Lau, T. K., Chang, A. M., and Hjelm, N. M. (1999) Rapid clearance of fetal DNA from maternal plasma. *Am. J. Hum. Genet.* **64,** 218–224.

6. Ariga, H., Ohto, H., Busch, M. P., et al. (2001) Kinetics of fetal cellular and cell-free DNA in the maternal circulation during and after pregnancy: implications for noninvasive prenatal diagnosis. *Transfusion* **41,** 1524–1530.

7. Smid, M., Galbiati, S., Vassallo, A., et al. (2003) No evidence of fetal DNA persistence in maternal plasma after pregnancy. *Hum. Genet.* **112,** 617–618.

8. Lo, Y. M. D., Leung, T. N., Tein, M. S., et al. (1999) Quantitative abnormalities of fetal DNA in maternal serum in preeclampsia. *Clin. Chem.* **45,** 184–188.

9. Farina, A., Sekizawa, A., Rizzo, N., et al. (2004) Cell-free fetal DNA (SRY locus) concentration in maternal plasma is directly correlated to the time elapsed from the onset of preeclampsia to the collection of blood. *Prenat. Diagn.* **24,** 293–297.

10. Lo, Y. M. D., Lau, T. K., Zhang, J., et al. (1999) Increased fetal DNA concentrations in the plasma of pregnant women carrying fetuses with trisomy 21. *Clin. Chem.* **45,** 1747–1751.

11. Wataganara, T., LeShane, E. S., Farina, A., et al. (2003) Maternal serum cell-free fetal DNA levels are increased in cases of trisomy 13 but not trisomy 18. *Hum. Genet.* **112,** 204–208.

12. Leung, T. N., Zhang, J., Lau, T. K., Hjelm, N. M., and Lo, Y. M. D. (1998) Maternal plasma fetal DNA as a marker for preterm labour. *Lancet* **352,** 1904–1905.

13. Lau, T. K., Lo, K. W., Chan, L. Y. S., Leung, T. Y., and Lo, Y. M. D. (2000) Cell-free fetal deoxyribonucleic acid in maternal circulation as a marker of fetal-maternal hemorrhage in patients undergoing external cephalic version near term. *Am J. Obstet. Gynecol.* **183,** 712–716.

14. Sekizawa, A., Jimbo, M., Saito, H., et al. (2002) Increased cell-free fetal DNA in plasma of two women with invasive placenta. *Clin. Chem.* **48,** 353–354.

15. Jimbo, M., Sekizawa, A., Sugito, Y., et al. (2003) Placenta increta: Postpartum monitoring of plasma cell-free fetal DNA. *Clin. Chem.* **49,** 1540–1541.

16. Sugito, Y., Sekizawa, A., Farina, A., et al. (2003) Relationship between severity of hyperemesis gravidarum and fetal DNA concentration in maternal plasma. *Clin. Chem.* **49,** 1667–1669.

17. Zhong, X. Y., Holzgreve, W., Li, J. C., Aydinli, K., and Hahn, S. (2000) High levels of fetal erythroblasts and fetal extracellular DNA in the peripheral blood of a pregnant woman with idiopathic polyhydramnios: case report. *Prenat. Diagn.* **20,** 838–841.

18. Lo, Y. M. D., Hjelm, N. M., Fidler, C., et al. (1998) Prenatal diagnosis of fetal RhD status by molecular analysis of maternal plasma. *N. Engl. J. Med.* **339,** 1734–1738.

19. Hromadnikova, I., Vechetova, L., Vesela, K., Benesova, B., Doucha, J., and Vik, R. (2005) Non-invasive fetal RHD and RHCE genotyping using real-time PCR testing of maternal plasma in RhD-negative pregnancies. *J. Histochem. Cytochem.* **53,** 301–305.

20. Chiu, R. W. K., Lau, T. K., Leung, T. N., et al. (2002) Prenatal exclusion of beta thalassaemia major by examination of maternal plasma. *Lancet* **360,** 998–1000.

21. Li, Y., Di Naro, E., Vitucci, A., Zimmermann, B., Holzgreve, W., and Hahn, S. (2005) Detection of paternally inherited fetal point mutations for beta-thalassemia using size-fractionated cell-free DNA in maternal plasma. *JAMA* **293,** 843–849.

22. Chiu, R. W. K., Lau, T. K., Cheung, P. T., Gong, Z. Q., Leung, T. N., and Lo, Y. M. D. (2002) Noninvasive prenatal exclusion of congenital adrenal hyperplasia by maternal plasma analysis: a feasibility study. *Clin. Chem.* **48,** 778–780.

23. Li, Y., Holzgreve, W., Page-Christiaens, G. C., Gille, J. J., and Hahn, S. (2004) Improved prenatal detection of a fetal point mutation for achondroplasia by the use of size-fractionated circulatory DNA in maternal plasma. *Prenat. Diagn.* **24,** 896–898.

24. Chiu, R. W. K. and Lo, Y. M. D. (2004) The biology and diagnostic applications of fetal DNA and RNA in maternal plasma. *Curr. Top. Dev. Biol.* **61,** 81–111.

25. Chiu, R. W. K., Poon, L. L. M., Lau, T. K., Leung, T. N., Wong, E. M. C., and Lo, Y. M. D. (2001) Effects of blood-processing protocols on fetal and total DNA quantification in maternal plasma. *Clin. Chem.* **47,** 1607–1613.

26. Lee, T., LeShane, E. S., Messerlian, G. M., et al. (2002) Down syndrome and cell-free fetal DNA in archived maternal serum. *Am. J. Obstet. Gynecol.* **187,** 1217–1221.

27. Lui, Y. Y. N., Chik, K. W., Chiu, R. W. K., Ho, C. Y., Lam, C. W., and Lo, Y. M. D. (2002) Predominant hematopoietic origin of cell-free DNA in plasma and

serum after sex-mismatched bone marrow transplantation. *Clin. Chem.* **48,** 421–427.

28. Thijssen, M. A., Swinkels, D. W., Ruers, T. J., and de Kok, J. B. (2002) Difference between free circulating plasma and serum DNA in patients with colorectal liver metastases. *Anticancer Res.* **22,** 421–425.
29. Jung, M., Klotzek, S., Lewandowski, M., Fleischhacker, M., and Jung, K. (2003) Changes in concentration of DNA in serum and plasma during storage of blood samples. *Clin. Chem.* **49,** 1028–1029.
30. Angert, R. M., LeShane, E. S., Lo, Y. M. D., Chan, L. Y. S., Delli-Bovi, L. C., and Bianchi, D. W. (2003) Fetal cell-free plasma DNA concentrations in maternal blood are stable 24 hours after collection: analysis of first- and third-trimester samples. *Clin. Chem.* **49,** 195–198.
31. Chiu, R. W. K. and Lo, Y. M. D. (2002) Preanalytical issues for circulating DNA analysis: technical aspects, semantics and need for standardization, in *Molecular Testing in Laboratory Medicine: Selections From Clinical Chemistry 1998–2001* (Bruns, D. E., Lo, Y. M. D., and Wittwer, C. T., eds.). AACC, Washington: pp. 309–310.
32. Lambert, N. C., Lo, Y. M. D., Erickson, T. D., et al. (2002) Male microchimerism in healthy women and women with scleroderma: cells or circulating DNA? A quantitative answer. *Blood* **100,** 2845–2851.
33. Invernizzi, P., Biondi, M. L., Battezzati, P. M., et al. (2002) Presence of fetal DNA in maternal plasma decades after pregnancy. *Hum. Genet.* **110,** 587–591.
34. Ng, E. K. O., Tsui, N. B. Y., Lam, N. Y., et al. (2002) Presence of filterable and nonfilterable mRNA in the plasma of cancer patients and healthy individuals. *Clin. Chem.* **48,** 1212–1217.
35. Chiu, R. W. K., Chan, L. Y. S., Lam, N. Y. L., et al. (2003) Quantitative analysis of circulating mitochondrial DNA in plasma. *Clin. Chem.* **49,** 719–726.
36. Ding, C., Chiu, R. W., Lau, T. K., et al. (2004) MS analysis of single-nucleotide differences in circulating nucleic acids: application to noninvasive prenatal diagnosis. *Proc. Natl. Acad. Sci. USA* **101,** 10,762–10,767.
37. Farina, A., Caramelli, E., Concu, M., et al. (2002) Testing normality of fetal DNA concentration in maternal plasma at 10–12 completed weeks' gestation: a preliminary approach to a new marker for genetic screening. *Prenat. Diagn.* **22,** 148–152.
38. Hromadnikova, I., Houbova, B., Hridelova, D., et al. (2003) Replicate real-time PCR testing of DNA in maternal plasma increases the sensitivity of non-invasive fetal sex determination. *Prenat. Diagn.* **23,** 235–238.

10

Clinical Applications of Plasma Epstein-Barr Virus DNA Analysis and Protocols for the Quantitative Analysis of the Size of Circulating Epstein-Barr Virus DNA

K. C. Allen Chan and Y. M. Dennis Lo

Summary

Nasopharyngeal carcinoma (NPC) is one of the most common cancers in Southern China. Epstein-Barr virus (EBV) infection is an important etiological factor of NPC. The fact that EBV genome is present in almost all NPC tissues renders it an ideal tumor marker for NPC. To date, quantitative analysis of plasma EBV DNA has been shown to be clinically useful for the detection, monitoring, and prognostication of NPC. In addition, the molecular nature of circulating EBV DNA has recently been identified as that of free DNA fragments; it is not contained inside intact virions. By quantitative size analysis, it is further demonstrated that more than 80% of these DNA fragments are less than 180 bp in size. In this chapter, the clinical applications of plasma EBV DNA analysis and the protocols for the quantitative analysis of the size of circulating EBV DNA will be discussed.

Key Words: EBV DNA; NPC; plasma DNA; size analysis.

1. Introduction

Epstein-Barr virus (EBV) has been implicated in the pathogenesis of several cancers including nasopharyngeal carcinoma (NPC) and some lymphomas (*1*). In Southern China, EBV genome could be demonstrated in almost all NPC tumor tissues (*2,3*). Although EBV infection is closely associated with NPC, initially, the idea of using circulating EBV DNA for the detection of NPC was not very popular among scientists because of the high prevalence of EBV infection in Southeast Asia. More than 95% of adults living in this area have been infected by EBV (*4*) and the virus would reside in some resting B-lymphocytes in a latent state (*1*). Therefore, scientists believed that EBV DNA would be detectable in a high proportion of healthy individuals. However, in 1998,

From: *Methods in Molecular Biology, vol. 336: Clinical Applications of PCR*
Edited by: Y. M. D. Lo, R. W. K. Chiu, and K. C. A. Chan © Humana Press Inc., Totowa, NJ

Mutirangura et al. showed that EBV DNA could be detected in the serum of 14 of 42 NPC patients but none of the 82 control subjects *(5)*. This observation suggests that the composition of the serum and the cellular compartments are significantly different. Later, our group developed a quantitative real-time polymerase chain reaction (PCR) assay for the detection of plasma EBV DNA, and we were able to detect EBV DNA in the plasma of 96% of NPC patients *(6)*. On the contrary, only 7% of healthy individuals showed positive results, and their plasma EBV DNA levels were much lower than the NPC patients *(6)*. Furthermore, the levels of plasma EBV DNA showed a strong correlation with the clinical stages of the patients *(6)*. The median concentration of the plasma EBV DNA of late-stage patients (stages III and IV) was almost eight times higher than that of the early-stage (stages I and II) patients *(6)*. These observations suggested that EBV DNA is tumor-derived and its concentration reflects the tumor burden. To date, quantitative analysis of circulating EBV DNA has been shown to be a very important tool for the detection *(6)*, monitoring *(7)*, and prognostication *(8)* of NPC and other EBV-associated malignancies *(9,10)*.

Despite a growing number of clinical applications of circulating EBV DNA analysis, the molecular nature of these DNA species was unclear. Regarding this issue, we used different approaches to study the nature of these EBV DNA molecules *(11)*. To discriminate whether the circulating EBV DNA molecules exist as DNA fragments or as a component of intact virions, we subjected the plasma of NPC patients to DNase digestion and ultracentrifugation *(11)*. First, we showed that viral particles were resistant to DNase digestion and are pelletable by ultracentrifugation *(11)*. In contrast, spiked extracted EBV DNA could no longer be detected in the plasma of NPC patients after DNase digestion and, after ultracentrifugation, most of the extracted EBV DNA remained in the supernatant *(11)*. The observation that circulating EBV DNA in NPC patients is susceptible to DNase digestion but not pelletable by ultracentrifugation suggests that plasma EBV DNA exists as DNA fragments instead of a part of an intact virion.

Furthermore, we developed a method for the quantitative analysis of the fragment size of circulating EBV DNA fragments. We have shown that 87% of circulating EBV DNA in NPC and lymphoma patients are below 181 bp in size *(11)*. This method has further been modified for the analysis of the size of circulating DNA in pregnant women by targeting a gene in the human genome (e.g., the *Leptin* gene) *(12)*. We showed that the circulating DNA in pregnant women is longer than that of the nonpregnant counterparts, and the fetal-derived DNA in maternal circulation is shorter than the maternally derived DNA *(12)*. The principle and protocols of the quantitative analysis of the size of circulating EBV DNA are discussed later.

For the measurement of the length of plasma EBV DNA molecules, a panel of 10 quantitative real-time PCR assays targeting a region encoding the EBV-

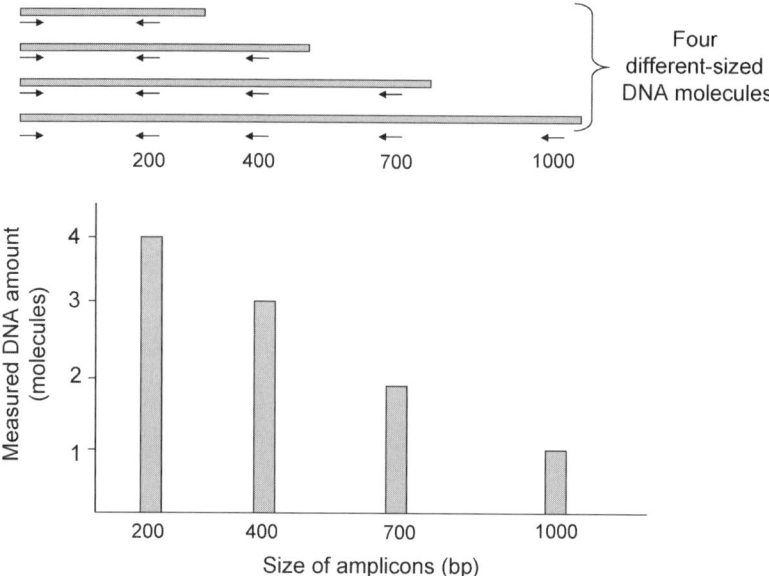

Fig. 1. Four different-sized DNA molecules are shown in the upper diagram. The arrow pointing to the right represents the forward primer and the arrows pointing to the left side represent different reverse primers. The DNA molecules can be amplified and detected by the real-time polymerase chain reaction only if both the forward and reverse primers can be annealed to the DNA molecules. With the increase of amplicon size, the number of DNA molecules detected by the system is decreased.

encoded RNAs (EBER) were developed *(11)*. The amplicon lengths of these PCR assays ranged from 82 to 1000 bp. These quantitative real-time assays could only detect DNA molecules longer than their respective amplicons. Therefore, for the same sample, the EBV DNA concentration measured by an assay with longer amplicon would be lower than that measured by an assay with shorter amplicon. The principle is illustrated in **Fig. 1**.

2. Materials

2.1. Sample Collection

Ethylenediamine tetraacetic acid (EDTA)-containing blood collection tubes for plasma collection.

2.2. DNA Extraction

1. Absolute ethanol.
2. QIAamp Blood Midi and Mini Kits (Qiagen, Hilden, Germany).
   a. Protease.
   b. Lysis buffer.

   c. Spin columns.

   d. Collection tubes.

   e. Wash buffer 1.

   f. Wash buffer 2.

3. Deionized water (dH$_2$O).

2.3. Real-Time PCR Assay

1. TaqMan PCR Reagent Kit (Perkin-Elmer):
   a. Ampli*Taq* Gold polymerase (5 U/mL).
   b. MgCl$_2$ (25 mM).
   c. dCTP, dATP, dGTP (200 μM each).
   d. dUTP (400 μM).
   e. Uracil-*N*-glycosylase (UNG) (1 U/μL).
   f. 10X TaqMan Buffer A.
   g. Forward primer, reverse primers and minor groove-binding (MGB) probe (Applied Biosystems) (**Table 1**).
   h. Molecular-grade dH$_2$O.
   i. Dimethylsulfoxide (DMSO) 100%.
2. Namalwa cell line (American Type Culture Collection no. CRL-1432).
3. Applied Biosystems ABI PRISM 7700 sequence detector.
4. Applied Biosystems MicroAmp Optical 96-well reaction plates.
5. Applied Biosystems optical caps (8 caps/strip).

3. Methods

3.1. Sample Types and Preparation

1. Collect peripheral blood into a tube containing EDTA as anti-coagulant.
2. Centrifuge the sample at 1600g for 10 min to separate plasma from cellular components.
3. Carefully transfer the supernatant into a 1.5-mL polypropylene tube without disturbing the cellular components.
4. Microcentrifuge the polypropylene tube at 16,000g for 10 min to ensure the removal of all cells.
5. Transfer the supernatant (cell-free plasma) into a clean polypropylene tube for subsequent DNA extraction (*see* **Note 1**).
6. Plasma samples used for size analysis of circulating DNA should be processed within 6 h of blood collection (*see* **Note 2**).
7. If the samples are to be used in different analyses carried out at different times, the plasma samples should be aliquoted into smaller fractions to avoid repeated freezing and thawing, or DNA should be extracted from the plasma samples for storage (*see* **Note 2**).

Table 1
List of Primers and Probes
for Real-Time Quantitative Polymerase Chain Reaction

Name	Sequence	Forward/ reverse/ probe	Amplicon size (bp)
EBER82F	5'-GAGAGGCTTCCCGCCTAGA-3'	Forward	82
EBER181F	5'-TACATCAAACAGGACAGCCGTT-3'	Forward	181
EBER294F	5'-TCCCGCAGTTCCACCTAAAC-3'	Forward	294
EBER385F	5'-GACTCTGCTTTCTGCCGTCTTC-3'	Forward	385
EBER493F	5'-CTACGCTGCCCTAGAGGTTTTG-3'	Forward	493
EBER584F	5'-ACACACCAACTATAGCAAACCCC-3'	Forward	584
EBER695F	5'-TCCCAGAGAGGGTAAAAGAGGG-3'	Forward	695
EBER781F	5'-CCCGCTACGTGCAGTGCT-3'	Forward	781
EBER891F	5'-TACAGCTAAATGCCCACCA-3'	Forward	891
EBER1000F	5'-GCAGAGGACATTGGGCAGGT-3'	Forward	1000
EBER-R	5'-AAATAGCGGACAAGCCGAATA-3'	Reverse	
EBERU3T	5'-(FAM)TCTCCCAGAGGGATTAGA(MGBNFQ)-3'	Probe	

EBER, Epstein-Barr virus-encoded RNAs; FAM, 6-carboxyfluorescein; MGBNFQ, minor groove-binding nonfluorogenic quencher.

3.2. DNA Extraction

For the quantitative analysis of plasma DNA, the same plasma sample would be subjected to multiple real-time quantitative PCRs of different amplicon sizes, and the number of analysis depends on the size range and the resolution of the size distribution. Therefore, a higher elution volume is necessary in order to give adequate material for the whole panel of analyses. Thus, we have been using the QIAamp DNA Midi Kit (Qiagen, Hilden, Germany) for DNA extraction from plasma for the size analysis of circulating DNA.

1. Follow the "blood and body fluid protocol" as per manufacturer's instructions.
2. 2 mL plasma is loaded to the column and eluted with 300 μL dH$_2$O *(11)*.

3.3. Real-Time PCR Assays

3.3.1. Design of TaqMan Assays

1. The 10 PCR assays shared a common forward primer and a common TaqMan probe, but consisted of 10 different reverse primers *(11)*.
2. The locations of the reverse primers are determined by the desired amplicon lengths.
3. Although the forward primer and the probe are used by all reactions, it would be more convenient if the melting temperatures of all the primers are approximately

the same. Otherwise, the optimization of the assays would be very difficult and the sensitivities of the assays would have wide variations (*see* **Note 3**). As the melting temperatures of all the primers were very close, the thermal cycles for all the assays were identical.

3.3.2. Establishing the Calibration Standards With Known Amounts of EBV DNA

1. For the quantification of circulating EBV DNA, a calibration curve must be established. DNA solutions with known quantities of EBV DNA could be prepared from the Namalwa cell line *(13)* (*see* **Note 4**).
2. Harvest Namalwa cells following tissue culture.
3. Extract the DNA from these cells using the QIAamp Mini Kit following the blood and body fluid protocol.
4. Measure the concentration of total DNA (*D* μg/mL) by spectrophotometry.
5. Translate the concentration of total DNA into EBV genome equivalents in each microliter of solution (*N* genomes/5 μL) by the following formula (*see* **Note 4**):

$$N = D \times 1000 \times 2 \text{ Y } 5/6.6$$

6. Dilute the DNA solution to obtain a standard concentration of 100,000 genomes/μL. Then, dilute the standard (100,000 genomes/μL) serially to obtain DNA solutions with concentrations of 10,000, 1000, 500, 100, 50, 25, 12.5, 6.25, 3.13, 1.57, and 0.78 genomes/μL.

3.3.3. Real-Time PCR for Assays With Differently Sized Amplicons

1. Prepare the PCR master mix according to the protocol in stated **Table 2** (*see* **Notes 5** and **6**).
2. Pipet 45 μL of master mix into each well of a MicroAmp Optical 96-well reaction plate.
3. Pipet 5 μL of each sample (including the calibration standards (*see* **Notes 7** and **8**) and plasma DNA samples of unknown concentrations) into each well in which the master mix has been added in duplicate.
4. Seal the 96-well reaction plate with optical caps.
5. Put the 96-well reaction plate in the reaction chamber of ABI 7700 sequence detector.
6. Input the sample names and types to the sequence detector program and start the program.
7. The real-time PCR reactions are carried out in an Applied Biosystem 7700 Sequence Detector with a thermal profile as stated in **Table 3**.

3.4. Presentation of Results

Real-time PCR assays with amplicons sized 82 to 382 bp were able to detect 1 copy of EBV DNA per reaction, whereas those with amplicons sized 493 to 1000 bp were able to detect 10 copies of EBV DNA per reaction. The plot of plasma EBV DNA concentrations in NPC patients is shown in **Fig. 2B**. As the

Table 2
Protocol for Preparation of Polymerase Chain Reaction Master Mix

	Volume for one reaction (μL)	Volume for N reactions (μL)
10X buffer A	5	$5 \times N$
MgCl$_2$ (25 mM)	8	$8 \times N$
dNTP mix^a	4	$4 \times N$
Forward primer (10 pmol/μL)	5	$5 \times N$
Reverse primer (10 pmol/μL)	5	$5 \times N$
MGB probe (2.5 pmol/μL)	1	$1 \times N$
Uracil-N-glycosylase (1 U/μL)	0.5	$0.5 \times N$
DMSO 100%	2.5	$2.5 \times N$
Ampli*Taq* Gold (5 U/μL)	0.4	$0.4 \times N$
H$_2$O	22.6	$22.6 \times N$
Total	45	$45 \times N$

^{a}dNTP mix: dATP, dCTP, dGTP at 50 μM; dUTP at 100 μM. MGB, minor groove-binding; DMSO, dimethylsulfoxide.

Table 3
Thermal Profile of Real-Time Polymerase Chain Reactions Carried Out in an Applied Biosystem 7700 Sequence Detector

50°C for 2 min
95°C for 1 min
95°C for 30 s ⎫
58°C for 1 min ⎬ 50 cycles
72°C for 1 min ⎭

EBV DNA concentrations differed by up to six orders of magnitude in different patients, the plots are compressed even with a logarithmic scale and the size distribution of plasma EBV DNA in these individuals could not be easily appreciated. To overcome this problem, the fractional concentration of each amplicon was calculated by dividing its absolute concentration by the absolute concentration determined by the assay with the shortest amplicon (82 bp). In this way, size distribution of DNA molecules from different patients can be summarized and compared. **Figure 2B** shows the plot of fractional concentrations against amplicon lengths. There is a sharp drop of DNA concentrations when the amplicon size is increased from 82 to 181 bp. On the other hand, the drop in fractional concentration beyond the amplicon size of 181 bp is gradual.

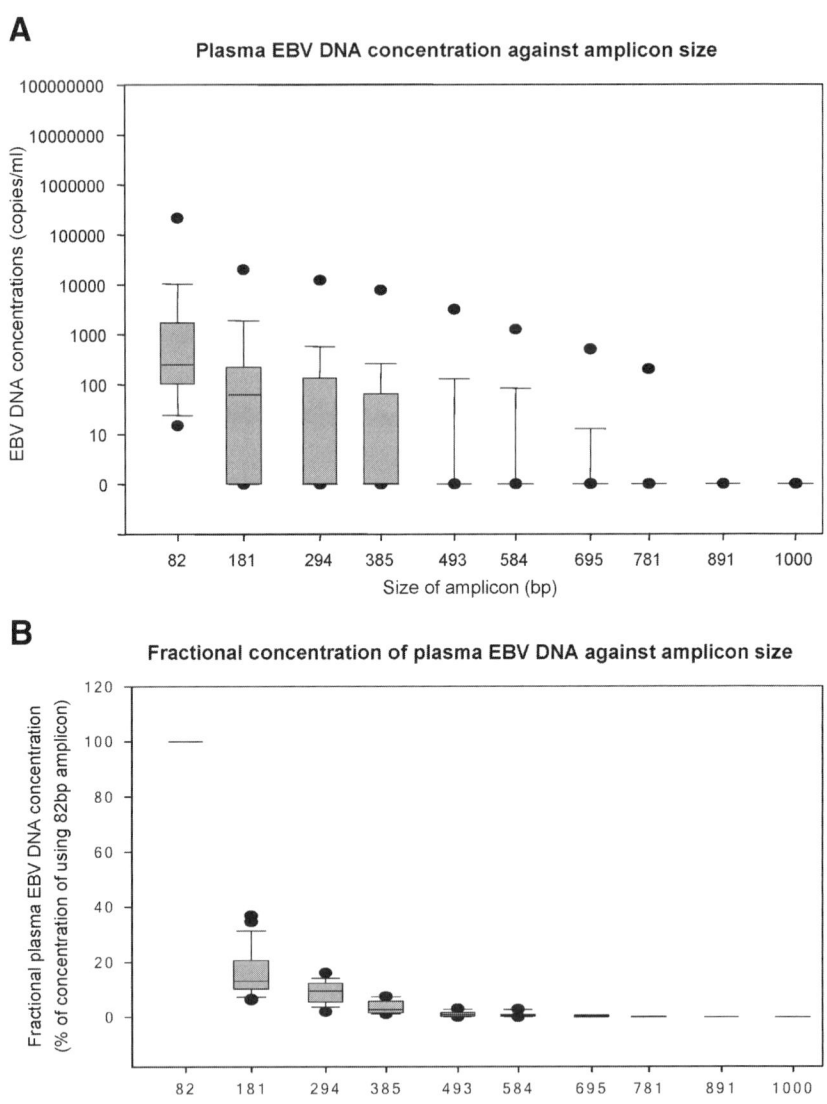

Fig. 2. (**A**) Plasma Epstein-Barr virus (EBV) DNA concentration against amplicon size in nasopharyngeal carcinoma patients. A wide dispersion of plasma EBV DNA concentrations for each amplicon size is observed. (**B**) Fractional concentration of EBV DNA against amplicon size. The plasma EBV DNA concentrations of each amplicon size were expressed as a fraction of the concentration measured by the 82 bp (shortest amplicon) assay of that patient. The size distribution of plasma EBV DNA is more easily appreciated with this presentation. A marked drop in fractional concentrations can be observed when the amplicon size is increased from 82 to 181 bp.

This finding implies that circulating EBV DNA in NPC patients mainly consists of short DNA fragments of less than 181 bp and that less than 1% of the plasma EBV DNA molecules are longer than 500 bp.

4. Notes

1. The described plasma processing protocol has been shown to effectively remove blood cells from plasma. Incomplete removal of blood cells would lead to an apparent increase in the size of circulating DNA.

2. The size of circulating DNA would be affected by sample types and preanalytical factors including the delay in separation of cellular components and repeated freezing and thawing of plasma samples. Blood clotting and a greater than 6-h delay in the separation of plasma from cellular components would lead to the release of high-molecular-weight DNA from blood cells into the fluid compartment *(14,15)*, and thus would result in an apparent lengthening of circulating genomic DNA in size analysis. On the other hand, repeated freezing and thawing of plasma samples would result in increased fragmentation of circulating DNA *(15)*. In contrast, no significant fragmentation of DNA would be observed after repeated freezing and thawing of extracted DNA for up to three times *(15)*.

3. During the optimization of assays, we also evaluated the differences between the use of a standard TaqMan probe and an MGB probe and found that the assays using MGB probe had lower detection limits when the amplicon lengths went beyond 400 bp. This may be due to the better binding of the MGB probe to the DNA template during the new strand synthesis.

4. Namalwa cell line is a diploid cell line containing two integrated EBV genomes/cell and it is incapable of producing free EBV. A conversion factor of 6.6 pg of DNA per diploid cell was used for copy number conversion. As the DNA molecules extracted from the Namalwa cell line are of high molecular weight, they could be detected by assays with different amplicon lengths.

5. All reactions should be run in duplicates and the average of two threshold cycle (C_T) values is used for the analysis for each sample of unknown quantity and for each calibration sample. Multiple water blanks are also required as negative controls and extra volume for compensation for pipetting errors (200 μL) is required when preparing the master mix. Therefore, the number of reactions N is equal to $2 \times$ (no. of unknown samples + 14).

6. The addition of UNG into the reaction mixture and the use of dUTP instead of dTTP is an anti-contamination strategy *(16)*. As dUTP is used in all PCR reactions carried out in our laboratory, all carryover PCR products would be destroyed by enzymatic digestion by the UNG during the 2-min incubation at 50°C (**step 1** of the thermal profile). In **step 2** of the thermal profile, the UNG is degraded and the AmpliTaq Gold polymerase is activated.

7. For circulating DNA size analysis, a separate calibration curve (PCR amplification) should be set up for each real-time PCR of different amplicon sizes. This size-specific calibration curve can minimize the effect of the different efficiencies of PCRs.

8. The same set of serially diluted Namalwa DNA should be used as the calibration material for different-sized PCR assays. This can minimize the lot-to-lot variation in the quantities of the calibration standards.

Acknowledgments

This work was supported by an Earmarked Research Grant (CUHK 4276/04M) from the Research Grants Council of the Hong Kong Special Administrative Region (China).

References

1. Rickinson, A. B. and Kieff, E. (1996) Epstein-Barr virus, in *Fields Virology*. Lippincott-Raven, Philadelphia: pp. 2397–2446.
2. Chen, C. L., Wen, W. N., Chen, J. Y., Hsu, M. M., and Hsu, H. C. (1993) Detection of Epstein-Barr virus genome in nasopharyngeal carcinoma by in situ DNA hybridization. *Intervirology* **36,** 91–98.
3. Dickens, P., Srivastava, G., Loke, S. L., Chan, C. W., and Liu, Y. T. (1992) Epstein-Barr virus DNA in nasopharyngeal carcinomas from Chinese patients in Hong Kong. *J. Clin. Pathol.* **45,** 396–397.
4. Dan, R. and Chang, R. S. (1990) A prospective study of primary Epstein-Barr virus infections among university students in Hong Kong. *Am. J. Trop. Med. Hyg.* **42,** 380–385.
5. Mutirangura, A., Pornthanakasem, W., Theamboonlers, A., et al. (1998) Epstein-Barr viral DNA in serum of patients with nasopharyngeal carcinoma. *Clin. Cancer Res.* **4,** 665–669.
6. Lo, Y. M. D., Chan, L. Y., Lo, K. W., et al. (1999) Quantitative analysis of cell-free Epstein-Barr virus DNA in plasma of patients with nasopharyngeal carcinoma. *Cancer Res.* **59,** 1188–1191.
7. Lo, Y. M. D., Chan, L. Y., Chan, A. T., et al. (1999) Quantitative and temporal correlation between circulating cell-free Epstein-Barr virus DNA and tumor recurrence in nasopharyngeal carcinoma. *Cancer Res.* **59,** 5452–5455.
8. Lo, Y. M. D., Chan, A. T., Chan, L. Y., et al. (2000) Molecular prognostication of nasopharyngeal carcinoma by quantitative analysis of circulating Epstein-Barr virus DNA. *Cancer Res.* **60,** 6878–6881.
9. Lo, Y. M. D., Chan, W. Y., Ng, E. K., et al. (2001) Circulating Epstein-Barr virus DNA in the serum of patients with gastric carcinoma. *Clin. Cancer Res.* **7,** 1856–1859.
10. Lei, K. I., Chan, L. Y. S., Chan, W. Y., Johnson, P. J., and Lo, Y. M. D. (2000) Quantitative analysis of circulating cell-free Epstein-Barr virus (EBV) DNA levels in patients with EBV-associated lymphoid malignancies. *Br. J. Haematol.* **111,** 239–246.
11. Chan, K. C. A., Zhang, J., Chan, A. T. C., et al. (2003) Molecular characterization of circulating EBV DNA in the plasma of nasopharyngeal carcinoma and lymphoma patients. *Cancer Res.* **63,** 2028–2032.

12. Chan, K. C. A., Zhang, J., Hui, A. B. Y., et al. (2004) Size distributions of maternal and fetal DNA in maternal plasma. *Clin. Chem.* **50,** 88–92.
13. Klein, G., Dombos, L., and Gothoskar, B. (1972) Sensitivity of Epstein-Barr virus (EBV) producer and non-producer human lymphoblastoid cell lines to superinfection with EB-virus. *Int. J. Cancer* **10,** 44–57.
14. Lui, Y. Y. N., Chik, K. W., Chiu, R. W. K., Ho, C. Y., Lam, C. W., and Lo, Y. M. D. (2002) Predominant hematopoietic origin of cell-free DNA in plasma and serum after sex-mismatched bone marrow transplantation. Clin. Chem. **48,** 421–427.
15. Chan, K. C. A., Yeung, S. W., Lui, W. B., Rainer, T. H., and Lo, Y. M. D. (2005) Effects of preanalytical factors on the molecular size of cell-free DNA in blood. *Clin. Chem.* **51,** 781–784.
16. Pang, J., Modlin, J., and Yolken, R. (1992) Use of modified nucleotides and uracil-DNA glycosylase (UNG) for the control of contamination in the PCR-based amplification of RNA. *Mol. Cell Probes* **6,** 251–256.

11

Molecular Analysis of Circulating RNA in Plasma

Nancy B. Y. Tsui, Enders K. O. Ng, and Y. M. Dennis Lo

Summary

Circulating RNA in plasma and serum is a newly developed area for molecular diagnosis. To date, increasing numbers of studies show that plasma and serum RNA could serve as both tumor- and fetal-specific markers for cancer detection and prenatal diagnosis, respectively. Recently, by introducing the highly sensitive one-step real-time quantitative reverse-transcription (RT)-polymerase chain reaction (PCR), these potentially valuable RNA species, which often only exist at low concentrations in plasma and serum, can now be readily detected and quantified. Following the successful quantification of *glyceraldehyde-3-phosphate dehydrogenase* (*GAPDH*) mRNA in plasma of normal individuals, several placenta-derived mRNA species, including the mRNA transcripts of *human placental lactogen* (*hPL*), the β-*subunit of human chorionic gonadotropin* (β*hCG*), and *corticotropin-releasing hormone* (*CRH*) were also quantified in plasma of pregnant women. These circulating placental RNA species have provided the fetal-polymorphism-independent markers for prenatal diagnosis. The achievement in detecting the placental RNA in maternal plasma represents a significant step toward the development of RNA markers for noninvasive prenatal gene expression profiling. This detection technique can be extended to access a wide range of disease conditions, such as cancer and trauma. The one-step, real-time quantitative RT-PCR is a highly sensitive and specific, yet practically simple, RNA detection technique. This powerful technology may allow the practical employment of circulating RNA in the high-throughput clinical screening and monitoring applications.

Key Words: Plasma RNA; circulating RNA; plasma RNA extraction; plasma RNA quantification; one-step real-time quantitative RT-PCR; *glyceraldehyde-3-phosphate dehydrogenase*; *GAPDH*; *human placental lactogen* mRNA; *hPL*.

1. Introduction

The discovery of circulating RNA in plasma has opened up new medical diagnostic possibilities. The utility of such RNA as molecular markers may allow the molecular detection, prognostication, diagnosis, and monitoring of different disease conditions, such as cancer, trauma, and prenatal disease. The

From: *Methods in Molecular Biology, vol. 336: Clinical Applications of PCR*
Edited by: Y. M. D. Lo, R. W. K. Chiu, and K. C. A. Chan © Humana Press Inc., Totowa, NJ

existence of circulating RNA in plasma was first reported in 1999 when Lo et al. observed the presence of the tumor-associated RNA in the plasma of patients with nasopharyngeal carcinoma *(1)*. In conjunction with the report by Kopreski et al. *(2)*, these publications represent the first demonstrations of reverse-transcription (RT)-polymerase chain reaction (PCR)-amplifiable tumor-derived RNA in the plasma of cancer patients. Since then, tumor-associated RNA targets that have been detected include mRNA of the telomerase components, *mammaglobin*, β*-catenin*, *CK19*, *huRNP-B1*, *Her2/neu*, and *CEA (3–8)*. In addition, the availability of such an approach has stimulated the first demonstration of the presence of fetal RNA in maternal plasma *(9)*.

Initially, the lability of RNA and the existence of ribonuclease in the plasma *(10)* make it surprising that circulating RNA should be detectable at all. However, recent demonstration by Ng and his colleagues showed that a significant portion of circulating RNA in plasma is associated with subcellular particles *(11)*. This particle-associated nature of circulating RNA may explain the surprising stability of such RNA in plasma *(12)*. As a result of this inherent stability of circulating RNA, the storage of plasma samples at –80°C will be sufficient to reduce degradation, and the development of robust RNA extraction from plasma becomes realistic.

For most of the published works on the detection of circulating RNA in plasma, techniques such as conventional RT-PCR that are relatively sensitive and specific have been used. However, this approach is limited to qualitative analysis and requires time-consuming post-PCR analysis, making the routine implementation of such methodology difficult. Real-time RT-PCR, on the other hand, which is a quantitative approach, is increasingly used for measuring the level of gene expression. This technique is based on the performance of RT-PCR in the presence of a dual-labeled fluorescent probe, which allows the fluorescence signals to be recorded and analyzed during PCR cycling *(13)*. Together with suitable instrumentation, the steps of amplification, detection, and quantification can be combined. Thus, this methodology runs as a closed-tube system and postamplification manipulation can be eliminated. Furthermore, it eliminates risk of contamination and minimizes hands-on time. When used appropriately, this technique is robust and highly suited for high-throughput screening application *(14,15)*.

With the use of the real-time quantitative RT-PCR, Dasi et al. have demonstrated that the concentration of *telomerase reverse transcriptase* mRNA in plasma was elevated in colorectal cancer patients *(16)*. This study employed relative quantification of mRNA, in which the results were expressed as a ratio of *telomerase* mRNA to *glyceraldehyde-3-phosphate dehydrogenase* *(GAPDH)* mRNA so as to normalize plasma RNA levels from different individuals. However, recent demonstration by Lo's group that the concentrations of particle-associated and nonparticle-associated *GAPDH* mRNA were significantly elevated in the plasma of cancer patients *(11)* indicates that caution should be taken when expressing or interpreting plasma RNA data as a ratio. As quantification of mRNA transcript can be either relative or absolute, to avoid the inappropriate normalization in relative quantification, Lo and his colleagues have further developed a real-time quantitative RT-PCR for absolute quantification of circulating RNA in plasma. The latter group has further demonstrated that placental mRNA transcripts in maternal plasma are readily detectable, and the placenta is an important source of fetal RNA in maternal plasma *(17)*. This study represents the first quantitative analysis of gene expression of the fetus by analyzing a plasma sample from the mother. In this first report, the absolute calibration curve was constructed by serial dilutions of high-performance liquid chromatography (HPLC)-purified single-stranded synthetic DNA oligonucleotides specifying the studied amplicon. Previous data have shown that such single stranded oligonucleotides reliably mimic the products of the reverse transcription step and produce calibration curves that are identical to those obtained using T7-transcribed RNA *(18)*. The use of such calibration methodology poses certain advantages: first, it provides absolute concentration of mRNA transcript in plasma regardless the utility of the housekeeping gene normalization; and second, as a result of the commercial availability of the synthetic oligonucleotides, it significantly simplifies the process of obtaining a calibration curve when compared with the labor-intensive preparation of calibration curve involving amplicon subcloning and in vitro transcription.

The two systems discussed as follows are based on our recent development of the absolute quantifications of circulating RNA in plasma. The *GAPDH* RT-PCR system aims to quantify the housekeeping gene transcript, *GAPDH*, present in the plasma of human subjects *(11,12,17)*. This can be used as a positive control system to verify the quality of extracted RNA from plasma. The second RT-PCR system aims to measure the concentration of placenta-expressed gene transcript, *human placental lactogen* (*hPL*), present in the plasma of pregnant women *(17)*. This *hPL* RT-PCR system can be potentially used in noninvasive prenatal monitoring.

2. Materials

2.1. Sample Collection

Ethylenediamine tetraacetic acid (EDTA)-containing blood collection tubes for plasma collection.

2.2. RNA Extraction

1. Trizol LS reagent (Invitrogen, Carlsbad, CA).
2. Chloroform.
3. Ethanol.
4. RNeasy Mini Kit (Qiagen, Hilden, Germany).

5. RNase-Free DNase Set (Qiagen).
6. RNase Away (Invitrogen).

2.3. Real-Time Quantitative RT-PCR

1. Primers (*see* **Note 1**).
   a. *GAPDH* mRNA:
      i. Forward primer: 5'-GAAGGTGAAGGTCGGAGT-3'.
      ii. Reverse primer: 5'-GAAGATGGTGATGGGATTTC-3'.
   b. *hPL* mRNA:
      i. Forward primer: 5'-CATGACTCCCAGACCTCCTTC-3'.
      ii. Reverse primer: 5'-TGCGGAGCAGCTCTAGATTG-3'.
2. Dual-labeled fluorescent probes (*see* **Note 1**):
   a. *GAPDH* mRNA:
      5'-(FAM)CAAGCTTCCCGTTCTCAGCC(TAMRA)-3'.
   b. *hPL* mRNA:
      5'-(FAM)TTCTGTTGCGTTTCCTCCATGTTGG(TAMRA)-3', where FAM
      is 6-carboxy-fluorescein; TAMRA is 6-carboxy-tetramethylrhodamine.
3. Calibrator:
   a. TaqMan® Human *GAPDH* Control Reagents (Applied Biosystems, Foster
      City, CA) (*see* **Note 2**).
   b. Synthetic DNA oligonucleotides specifying the *hPL* amplicon, HPLC-
      purified (Proligos, Singapore) (*see* **Note 3**):
      5'-TGCGGAGCAGCTCTAGATTGGATTTCTGTTGCGTTTCCTCC
      ATGTTGGAGGGTGTCGGAATAGAGTCTGAGAAGCAGAAGGAGGTCT
      GGGAGTCATGC-3'.
4. RNase-free water.
5. EZ r*Tth* RNA PCR reagent set (Applied Biosystems).
6. ABI Prism 7700 Sequence Detector (Applied Biosystems).

3. Methods
3.1. Sample Collection

1. Collect blood samples in EDTA tubes. To ensure a sufficient amount of plasma
   for RNA analysis, at least 4 mL of blood should be taken for each sample (*see*
   Note 4). To guarantee a good quality of RNA, the blood samples should be
   processed as soon as possible after venesection. If the processing procedures
   cannot take place immediately, the blood samples should be stored with extra
   care (*see* **Note 5**).
2. Centrifuge the blood samples at 1600*g* for 10 min at 4°C.
3. Transfer plasma into new tubes.
4. Perform the second round centrifugation at 16,000*g* for 10 min at 4°C (*see* **Note 6**).
5. Add 2 mL of Trizol LS reagent to 1.6 mL of plasma (*see* **Note 7**) and vortex
   vigorously.
6. Store the Trizol-plasma mixture at –80°C until RNA extraction.

3.2. RNA Extraction

The RNA extraction should be performed in a clean area to minimize RNase contamination. RNase Away can be used to clean lab bench and equipment before each extraction.

In this protocol, total plasma RNA is extracted.

1. Thaw the Trizol-plasma mixture immediately before extraction.
2. Add 0.4 mL of chloroform to the mixture (2 mL Trizol LS reagent and 1.6 mL plasma) and mix vigorously.
3. Centrifuge the mixture at 12,000*g* for 15 min at 4°C.
4. Transfer the upper aqueous layer into new tubes. Add one volume of 70% ethanol to one volume of the aqueous layer.
5. Apply the mixture to an RNeasy Mini column and wash the column according to the manufacturer's recommendations.
6. Perform the "On-Column DNase Digestion" with the RNase-Free DNase Set according to the manufacturer's recommendations (*see* **Note 8**).
7. To elute RNA, add 30 µL of RNase-free water into the RNeasy column and incubate for 5 min at room temperature. Centrifuge the column for 1 min at 16,000*g*.
8. Repeat the elution step (**step 7**) by using the first eluate.
9. Store the extracted RNA at –80°C.

3.3. Real-Time Quantitative RT-PCR

Plasma RNA is quantified by using one-step real-time quantitative RT-PCR *(13)*. In this method, the r*Tth* (*Thermus thermophilus*) DNA polymerase functions both as a reverse transcriptase and a DNA polymerase *(19)* (*see* **Note 9**). In this protocol, the quantification of *GAPDH (11)* and *hPL (17)* mRNA is described as follows.

3.3.1. Quantification of GAPDH mRNA

1. Prepare a calibration curve for *GAPDH* RT-PCR system with the use of the human control RNA supplied in the TaqMan Human *GAPDH* Control Reagents. The control RNA is serially diluted into concentrations ranging from 15 ng to 1.5 pg (*see* **Note 2**).
2. Set up the RT-PCR reaction mixture for the *GAPDH* mRNA according to **Table 1**.
3. Add 3 µL of sample RNA, control RNA (for calibration curve) or RNase-free water (for negative blanks) into the reaction mixture.
4. Perform the real-time *GAPDH* RT-PCR reactions in the ABI Prism 7700 Sequence Detector with cycling conditions show in **Table 2**.

3.3.2. Quantification of hPL mRNA

1. Prepare a calibration curve by serially diluting the synthetic DNA oligonucleotides specifying the *hPL* amplicon with concentrations ranging from 2.5×10^7 copies/µL to 25 copies/µL (*see* **Note 3**).

Table 1
Compositions of Polymerase Chain Reaction Mix
for Amplification of *GAPDH* mRNA

Component	Volume for one reaction (μL)	Final concentration
5X TaqMan EZ Buffer	10	1X
Mn(OAc)$_2$ (25 mM)	6	3 mM
dATP (10 mM)	1.5	300 μM
dCTP (10 mM)	1.5	300 μM
dGTP (10 mM)	1.5	300 μM
dUTP (20 mM)	1.5	600 μM
Forward primer (10 μM)	1	200 nM
Reverse primer (10 μM)	1	200 nM
Probe (5 μM)	1	100 nM
r*Tth* DNA polymerase (2.5 U/mL)	2	0.1 U/μL
AmpErase UNG (1 U/μL)	0.5	0.01 U/μL
RNase-free water	19.5	–
Total volume	47	

GAPDH, glyceraldehyde-3-phosphate dehydrogenase; UNG, uracil-*N*-glycosylase.

Table 2
Cycling Profile for Amplification of *GAPDH* mRNA

Step		Temperature	Time
UNG treatment		50°C	2 min
Reverse transcription		60°C	30 min
Deactivation of UNG		95°C	5 min
40 Cycles	Denaturation	94°C	20 s
	Annealing/extension	62°C	1 min

GAPDH, glyceraldehyde-3-phosphate dehydrogenase; UNG, uracil-*N*-glycosylase.

2. Set up the RT-PCR reaction mixture for *hPL* mRNA according to **Table 3**.
3. Add 6 μL of sample RNA, synthetic DNA oligonucleotides (for calibration curve) or RNase-free water (for negative blanks) into the reaction mixture.
4. Perform the real-time *hPL* RT-PCR reactions in the ABI Prism 7700 Sequence Detector with cycling conditions as shown in **Table 4**.

Table 3
Compositions of Polymerase Chain Reaction Mix
for Amplification of *hPL* mRNA

Component	Volume for one reaction (µL)	Final concentration
5X TaqMan EZ Buffer	10	1X
Mn(OAc)$_2$ (25 mM)	6	3 mM
dATP (10 mM)	1.5	300 µM
dCTP (10 mM)	1.5	300 µM
dGTP (10 mM)	1.5	300 µM
dUTP (20 mM)	1.5	600 µM
Forward primer (10 µM)	1.5	300 nM
Reverse primer (10 µM)	1.5	300 nM
Probe (5 µM)	1	100 nM
r*Tth* DNA polymerase (2.5 U/µL)	2	0.1 U/µL
AmpErase UNG (1 U/µL)	0.5	0.01 U/µL
RNase-free water	15.5	–
Total volume	44	

hPL; *human placental lactogen*; UNG, uracil-*N*-glycosylase.

Table 4
Cycling Profile for Amplification of hPL mRNA

Step		Temperature	Time
UNG treatment		50°C	2 min
Reverse transcription		60°C	30 min
Deactivation of UNG		95°C	5 min
	Denaturation	94°C	20 s
45 Cycles			
	Annealing/extension	56°C	1 min

hPL; *human placental lactogen*; UNG, uracil-*N*-glycosylase.

3.3.3. Data Analysis

Amplification data were analyzed and stored by the Sequence Detection System Software (v1.9; Applied Biosystems). The *GAPDH* mRNA concentrations are expressed as picogram of total RNA per milliliter of plasma (pg/mL) whereas the *hPL* mRNA concentrations are expressed as copies of *hPL* transcripts per milliliter of plasma (copies/mL). The calculation is shown as follows:

$$C = Q \times \frac{V_{RNA}}{V_{Plasma}}$$

in which C represents the target RNA concentration in plasma (pg/mL for *GAPDH*; copies/mL for *hPL*), Q represents the target RNA concentrations of the extracted RNA samples (pg/μL for *GAPDH*, copies/μL for *hPL*), which is determined by a sequence detector in a PCR, V_{RNA} represents the total volume of RNA obtained after extraction (typically 30 μL), V_{Plasma} represents the volume of plasma used for extraction (typically 1.6 mL).

The validations of both the *GAPDH* mRNA *(11)* and *hPL* mRNA *(17)* real-time RT-PCR systems are described in previous publications.

4. Notes

1. Primers and probes are designed with the use of *Primer Express*® Software v2.0 (Applied Biosystems). Certain precautions for the design are listed as follows:
   a. The amplicon length should be less than 100 bp, ideally no longer than 80 bp. Short amplicon length is preferable for several reasons: (1) the synthetic oligonucleotide which is used as a calibration curve (*see* **Note 3**) is commercially available with size only up to 100 nucleotides. Thus, the amplicon length is limited to 100 bp; and (2) amplification with shorter amplicon length is more efficient than that involving longer amplicon length *(18)*.
   b. Primers for amplification should ideally be intron-spanning in order to minimize any amplification of contaminating genomic DNA. The primer/probe locations of *GAPDH* and *hPL* mRNA are showed in **Fig. 1**. By using intron-spanning primers, intron-containing amplification products from the contaminating genomic DNA are too long to be amplified and thus the amplification will be cDNA-specific. If the intron/exon junctions are unknown, or when targeting an intron-less gene, it is necessary to perform DNase treatment in RNA samples (*see* **Note 8**).
   c. To avoid false-positive results arising from co-amplification of genes with high homology, it is necessary to perform a BLASTN search with the primer and probe sequences against the NCBI GenBank. The result of such search will provide information regarding the specificity of the amplification.
2. For the *GAPDH* RT-PCR system, the use of synthetic oligonucleotides as the calibration curve (*see* **Note 3**) is impossible because of the 226-bp long *GAPDH* amplicon. In this situation, the Human Control RNA is used instead. The manufacturer estimates that 1 pg of this control RNA contains approx 100 copies of *GAPDH* transcript.
3. Generally, in vitro-transcribed RNA is used as a calibration curve for an absolute RNA quantification. This in vitro-transcribed RNA is usually generated by subcloning the amplicon behind a T7 RNA polymerase promoter in a plasmid vector. However, this procedure is labor-intensive and time-consuming, and is unsuitable when a large number of RT-PCR systems need to be constructed in a

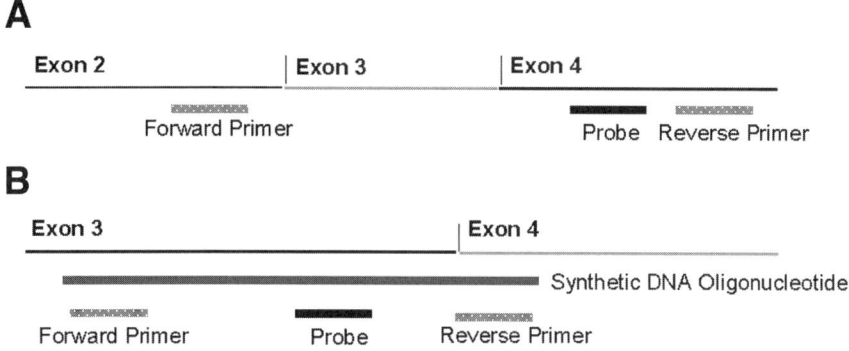

Fig. 1. **(A)** Locations of primers and probe for *glyceraldehyde-3-phosphate dehydrogenase* (*GAPDH*) mRNA. **(B)** Locations primers, probe and synthetic DNA oligonucleotide for *human placental lactogen* (*hPL*) mRNA.

short period of time. An alternative method that is increasingly used is to construct a calibration curve with the use of synthetic single-stranded oligonucleotides specifying the various amplicons. Previous data have shown that such single stranded oligonucleotides reliably mimic the products of the reverse transcription step and produce calibration curves that are identical to those obtained using T7-transcribed RNA *(18)*. The synthetic oligonucleotide for hPL mRNA is showed in **Fig. 1B**.

4. Our studies have revealed that 1.6 mL of plasma is the minimal sample volume for robust plasma RNA detection for many RNA targets. Thus, we strongly suggest that at least 4 mL of blood samples should be collected. For the *GAPDH* RT-PCR system, the minimal plasma sample volume is 0.6 mL.

5. When the blood samples are left unprocessed, their corresponding plasma and serum RNA concentrations will fluctuate over time *(12)*. This artifactual fluctuation may be due to several factors, such as release of RNA from necrotic and/or apoptotic blood cells, and the stability of the original and the newly released RNA. To guarantee a reliable plasma and serum RNA concentrations, we recommend all blood samples, including EDTA blood and clotted blood, be stored at 4°C and be processed within 6 h.

6. Based on our previous report that blood processing by different protocols would affect the result of plasma DNA quantitation *(20,21)*, our current protocol involves an initial 1600*g* centrifugation of blood samples followed by a second centrifugation at 16,000*g*. This protocol has been shown to minimize the contribution of RNA and DNA derived from residual cells in plasma.

7. A recent study has shown that RNA in frozen plain plasma would degrade with time *(22)*. Therefore, to preserve the integrity of RNA, Trizol LS reagent should be added to the plasma before storage if the sample is to be archived *(22)*. Based on our studies (unpublished), if RNA extraction would be performed within 2 wk, it is acceptable to store plasma at the following conditions: (1) –80°C with or with-

out Trizol LS reagent, or (2) –20°C only with Trizol LS reagent. If the plasma samples are stored without Trizol LS reagent, it should be added immediately before the addition of chloroform during RNA extraction. However, we observed significant drops in RNA concentration when plasma was stored at –20°C without Trizol LS reagent even when stored even for 2 wk.

8. Any significant DNA contamination in RNA samples will result in inaccurate RNA quantification particularly when non-intron-spanning primers are used. DNase I treatment must be carried out to eliminate any residual genomic DNA left during RNA extraction. In addition, it is necessary to test the RNA samples to ensure they are negative for DNA by substituting the r*Tth* polymerase with the Ampli *Taq* Gold enzyme (Applied Biosystems). For buffy coat or whole blood samples, the "On-Column DNase Digestion" appears not to completely remove all genomic DNA. Other stronger DNase digestion method, such as deoxyribonuclease I from Invitrogen, can be used instead.

9. The use of one-step one-enzyme RT-PCR with r*Tth* polymerase has several advantages over the two-enzyme RT-PCR:
   a. It has been reported that *Tth* polymerase is more resistant to inhibitors present in biological specimens than *Taq* polymerase *(23)*.
   b. The thermostable nature of r*Tth* polymerase allows the RT to be performed at a high temperature of 60°C. This minimizes secondary structures present in RNA as well as reduces primer dimers and nonspecific bindings during the RT reaction.
   c. In two-enzyme RT-PCR, the PCR product carryover prevention enzyme, uracil-*N*-glycosylase (UNG), cannot be used because the substitution of dUTP for dTTP in the RT reaction makes the nascent cDNA a substrate for UNG. On the other hand, the thermostable r*Tth* polymerase permits the use of UNG becasue the RT temperature of 60°C inactivates the UNG activity. As the RT reaction and PCR are carried out together in a single tube, this reduces both the hands-on time and the risk of contamination.

Acknowledgments

This work was supported by a Earmarked Research Grant (CUHK 4474/03M) from the Research Grants Council of the Hong Kong Special Administrative Region (China).

References

1. Lo, K. W., Lo, Y. M. D., Leung, S. F., et al. (1999) Analysis of cell-free Epstein-Barr virus associated RNA in the plasma of patients with nasopharyngeal carcinoma. *Clin. Chem.* **45,** 1292–1294.
2. Kopreski, M. S., Benko, F. A., Kwak, L. W., and Gocke, C. D. (1999) Detection of tumor messenger RNA in the serum of patients with malignant melanoma. *Clin. Cancer Res.* **5,** 1961–1965.
3. Chen, X. Q., Bonnefoi, H., Pelte, M. F., et al. (2000) Telomerase RNA as a detection marker in the serum of breast cancer patients. *Clin. Cancer Res.* **6,** 3823–3826.

4. Wong, S. C. C., Lo, E. S. F., Cheung, M. T., et al. (2004) Quantification of plasma beta-catenin mRNA in colorectal cancer and adenoma patients. *Clin. Cancer Res.* **10,** 1613–1617.
5. Gal, S., Fidler, C., Lo, Y. M. D., et al. (2001) Detection of mammaglobin mRNA in the plasma of breast cancer patients. *Ann. N. Y. Acad. Sci.* **945,** 192–194.
6. Novakovic, S., Hocevar, M., Zgajnar, J., Besic, N., and Stegel, V. (2004) Detection of telomerase RNA in the plasma of patients with breast cancer, malignant melanoma or thyroid cancer. *Oncol. Rep.* **11,** 245–252.
7. Fleischhacker, M., Beinert, T., Ermitsch, M., et al. (2001) Detection of amplifiable messenger RNA in the serum of patients with lung cancer. *Ann. N. Y. Acad. Sci.* **945,** 179–188.
8. Silva, J. M., Rodriguez, R., Garcia, J. M., et al. (2002) Detection of epithelial tumour RNA in the plasma of colon cancer patients is associated with advanced stages and circulating tumour cells. *Gut* **50,** 530–534.
9. Poon, L. L. M., Leung, T. N., Lau, T. K., and Lo, Y. M.D. (2000) Presence of fetal RNA in maternal plasma. *Clin. Chem.* **46,** 1832–1834.
10. Reddi, K. K. and Holland, J. F. (1976) Elevated serum ribonuclease in patients with pancreatic cancer. *Proc. Natl. Acad. Sci. USA* **73,** 2308–2310.
11. Ng, E. K. O., Tsui, N. B. Y., Lam, N. Y., et al. (2002) Presence of filterable and nonfilterable mRNA in the plasma of cancer patients and healthy individuals. *Clin. Chem.* **48,** 1212–1217.
12. Tsui, N. B. Y., Ng, E. K. O., and Lo, Y. M. D. (2002) Stability of endogenous and added RNA in blood specimens, serum, and plasma. *Clin. Chem.* **48,** 1647–1653.
13. Gibson, U. E., Heid, C. A., and Williams, P. M. (1996) A novel method for real time quantitative RT-PCR. *Genome Res.* **6,** 995–1001.
14. Huggett, J. Dheda, K., Bustin, S., and Zumia, A. (2005) Real-time RT-PCR normalisation; strategies and considerations. *Genes Immun.* **6,** 279–284.
15. Bustin, S. A. and Nolan, T. (2004) Pitfalls of quantitative real-time reverse-transcription polymerase chain reaction. *J. Biomol. Tech.* **15,** 155–166.
16. Dasi, F., Lledo, S., Garcia-Granero, E., et al. (2001) Real-time quantification in plasma of human telomerase reverse transcriptase (hTERT) mRNA: a simple blood test to monitor disease in cancer patients. *Lab. Invest.* **81,** 767–769.
17. Ng, E. K. O., Tsui, N. B. Y., Lau, T. K., et al. (2003) mRNA of placental origin is readily detectable in maternal plasma. *Proc. Natl. Acad. Sci. USA* **100,** 4748–4753.
18. Bustin, S. A. (2000) Absolute quantification of mRNA using real-time reverse transcription polymerase chain reaction assays. *J. Mol. Endocrinol.* **25,** 169–193.
19. Myers, T. W. and Gelfand, D. H. (1991) Reverse transcription and DNA amplification by a Thermus thermophilus DNA polymerase. *Biochemistry* **30,** 7661–7666.
20. Chiu, R. W. K., Poon, L. L. M., Lau, T. K., Leung, T. N., Wong, E. M. C., and Lo, Y. M. D. (2001) Effects of blood-processing protocols on fetal and total DNA quantification in maternal plasma. *Clin. Chem.* **47,** 1607–1613.
21. Chan, K.C. A., Yeung, S. W., Lui, W. B., Rainer, T. H., and Lo, Y. M. D. (2005) Effects of preanalytical factors on the molecular size of cell-free DNA in blood. *Clin. Chem.* **51,** 781–784.

22. Wong S. C. C., Lo E. S. F., and Cheung M. T. (2004) An optimised protocol for the extraction of non-viral mRNA from human plasma frozen for three years. *J. Clin. Pathol.* **57,** 766–768.
23. Poddar, S. K., Sawyer, M. H., and Connor, J. D. (1998) Effect of inhibitors in clinical specimens on Taq and Tth DNA polymerase-based PCR amplification of influenza A virus. *J. Med. Microbiol.* **47,** 1131–1135.

12

Molecular Analysis of Mitochondrial DNA Point Mutations by Polymerase Chain Reaction

Lee-Jun C. Wong, Bryan R. Cobb, and Tian-Jian Chen

Summary

Mitochondrial respiratory chain disorders are clinically and genetically heterogeneous. There are several mitochondrial DNA (mtDNA) point mutations responsible for common mitochondrial diseases such as mitochondrial encephalopathy, lactic acidosis, stroke-like events, myoclonic epilepsy and ragged red fibers, neuropathy, ataxia, retinitis pigmentosa, and Leber's hereditary optic neuropathy. As a result of the clinical overlap, it is usually necessary to analyze more than one mutation for a patient suspected of a mitochondrial disorder. Molecular diagnosis is often performed using polymerase chain reaction (PCR)/restriction fragment length polymorphism (RFLP) analysis of the most likely point mutations. However, this method is time-consuming and often produces problems associated with incomplete restriction enzyme digestion. In addition, PCR/RFLP analysis may not be able to detect a low percentage of heteroplasmy. For a more effective method of diagnosing mtDNA disorders, we have developed a multiplex PCR/allele-specific oligonucleotide (ASO) dot blot hybridization method to simultaneously analyze 11 point mutations. The PCR products from a DNA sample containing a homoplasmic wild-type or mutant mtDNA sequence will hybridize to either the wild-type or the mutant ASO probe. The PCR products of a heteroplasmic DNA sample will hybridize to both wild-type and mutant ASO probes. This PCR/ASO method allows the detection of low percentage mutant heteroplasmy.

Key Words: Mitochondrial disorders; mtDNA mutation; heteroplasmy; oxidative phosphorylation (OXPHOS) disease; MELAS; MERRF; NARP; LHON; mitochondrial dysfunction; Leigh's syndrome; respiratory chain disorders.

1. Introduction

Mitochondrial respiratory chain disorders are a group of clinically and genetically heterogeneous diseases *(1–3)*. Diagnosis is difficult because of the broad spectrum of phenotypic manifestations and the considerable clinical overlap between these disorders. The heteroplasmic nature of pathogenic point

From: *Methods in Molecular Biology, vol. 336: Clinical Applications of PCR*
Edited by: Y. M. D. Lo, R. W. K. Chiu, and K. C. A. Chan © Humana Press Inc., Totowa, NJ

Table 1
Primer Sequences

Primer	Primer site	Forward (F)/ Reverse (R)	Sequence (5'–3')	PCR product size
mtF3130	3130–3149	F	AGGACAAGAGAAATAAGGCC	651 bp
mtR3758	3758–3780	R	AGTAGAATGATGGCTAGGGTGAC	
mtF8278	8278–8297	F	CTACCCCTCTAGAGCCCAC	216 bp
mtR8475	8475–8493	R	TTTATCCCGTTTGGTCAGG	
mtF8768	8768–8785	F	CAACTAACCTCCTCGGAC	432 bp
mtR9199	9199–9180	R	TGTCGTGCAGGTAGAGGCTT	
mtF11688	11688–11705	F	CCGGCGCAGTCATTCTCA	672 bp
mtR12360	12360–12342	R	GGTTATAGTAGTGTGCATG	
mtF14437	14437–14455	F	AGGATACTCCTCAATAGCC	748 bp
mtR15185	15202–15185	R	GGCGGATAGTAAGTTTGT	

mutations in mitochondrial DNA (mtDNA) and a threshold effect further complicate the diagnosis. Currently, the most commonly known syndromes associated with specific mtDNA point mutations are myopathy, encephalopathy, lactic acidosis, and stroke-like episodes (MELAS); myoclonic epilepsy and ragged red fibers (MERRF); neuropathy, ataxia, and retinitis pigmentosa (NARP); and Leber's hereditary optic neuropathy (LHON) *(1–3)*. A total of 11 point mutations can be analyzed using two single polymerase chain reaction (PCR) assays and one multiplex PCR containing three pairs of primers. Detection of these point mutations is achieved by amplification of the DNA regions containing the point mutations followed by allele-specific oligonucleotide (ASO) hybridization *(4)*.

2. Materials

1. Extracted genomic DNA from patient's tissues *(5,6)*.
2. PCR primers *(4,7)*. Stock solutions are 100 μ*M* stored in a –20°C freezer. Working solutions are 10 μ*M*. Primers are listed in **Table 1**.
3. ASO probes *(4,7)*. Stock solutions are 100 μ*M* and stored at –20°C. Working solutions are 10 μ*M*. ASO probes are listed in **Table 2**. The nucleotide positions where the mutations occur are in bold *(see* **Note 1***)*.
4. PCR reagents:
   a. 10X PCR buffer II (Applied Biosystems, Inc. [ABI]).
   b. dNTP 8 m*M* solution.
   c. 25 m*M* MgCl$_2$.
   d. Ampli *Taq* or *Taq* Gold DNA polymerase (ABI).
5. Church prehybridization and hybridization buffer: 0.5 *M* Na$_2$HPO$_4$, pH 7.2, 1 m*M* ethylenediamine tetraacetic acid (EDTA), and 7% sodium dodecyl sulfate (SDS).

Table 2
Sequences of Allele-Specific Oligonucleotide Probes

Disease	Mutation	Probe sequence (5' to 3')	Sense(S)/ Antisense(A)
MELAS	A3243	TTACCGGGCTCTGCCATCT	A
	A3243G	TTACCGGGCCCTGCCATCT	A
MELAS	T3271	ACTTAAAACTTTACAGTCA	S
	T3271C	ACTTAAAACCTTACAGTCA	S
LHON	G3460	GAGTTTTATGGCGTCAGCGAA	A
	G3460A	GAGTTTTATGGTGTCAGCGAA	A
MERRF	A8344	GAGGTGTTGGTTCTCTTAAT	A
	A8344G	GAGGTGTTGGCTCTCTTAAT	A
MERRF	T8356	AACACCTCTTTACAGTGAA	S
	T8356C	AACACCTCTCTACAGTGAA	S
CARDIO-	G8363	TTTACAGTGAAATGCCCC	S
MYOPATHY	G8363A	TTTACAGTAAAATGCCCC	S
NARP	T8993	CAATAGCCCTGGCCGTACG	S
	T8993C	CAATAGCCCCGGCCGTACG	S
	T8993G	CAATAGCCCGGGCCGTACG	S
LHON	G11778	ATTATGATGCGACTGTGAG	A
	G11778A	ATTATGATGTGACTGTGAG	A
LHON	G14459	ATAGCCATCGCTGTAGTATA	S
	G14459A	ATAGCCATCACTGTAGTAT	S
LHON	T14484	AGACAACCATCATTCCC	S
	T14484C	AGACAACCACCATTCCC	S

MELAS, mitochondrial encephalopathy, lactic acidosis, stroke-like events; LHON, Leber's hereditary optic neuropathy; MERRF, myoclonic epilepsy and ragged red fibers; NARP, neuropathy, ataxia, retinitis pigmentosa.

6. 20X SSC: 3 M NaCl and 0.3 M trisodium citrate (Invitrogen, cat. no. 15557-044).
7. Church wash buffer: 40 mM phosphate buffer, pH 7.2, and 1% SDS.
8. 10 mg/mL Sheared salmon sperm DNA.
9. Agarose.
10. Agarose gel electrophoresis buffer (1X TAE): 40 mM Tris-acetate and 1 mM EDTA, pH 8.0, or 50X Tris Acetate-EDTA (Quality Biological, Inc., cat. no. 351-008-130).
11. 3X agarose gel loading dye: 0.125% bromophenol blue, 0.125% xylene cyanol, and 7.5% of Ficoll 400.
12. T4 polynucleotide kinase (10,000 U/mL) from New England Biolabs.
13. T4 polynucleotide kinase buffer: 70 mM Tris-HCl, pH 7.6, 10 mM MgCl$_2$, and 5 mM dithiothreitol (DTT).
14. γ-32P ATP (4500 Ci/mmol, 10 μCi/μL).
15. Zeta-Probe GT membrane (Bio-Rad, cat. no. 162-0196).

16. Hybridization oven.
17. Vacuum oven.

3. Methods

Mutant and normal controls as well as a water (no DNA template) control are always included in the PCR reactions. All control specimens are tested and analyzed in the same manner as the patient's samples. The DNA template for each positive control can be the patient's genomic DNA or PCR product containing the mutation. If the positive mutant control is not available, a synthetic control should be used *(8)*.

3.1. Preparation of Synthetic Positive Controls

1. Synthetic positive controls are generated by PCR using the ASO as one of the primers, which will produce a mutation-containing synthetic control PCR product *(8)*. For example, to generate a T3271C-containing positive control, the PCR reaction in **Table 3** is carried out in a total volume of 100 μL.
2. PCR conditions:
    a. 94°C, 4 min.
    b. 94°C, 20 s; 45°C, 20 s; 72°C, 30 s, for 30 cycles.
    c. 72°C, 5 min.
    d. 4°C soak.

3.2. PCR Setup (4)

Three sets of PCR reactions are set up. The first set is a single PCR for A3243G, T3271C, and G3460A mutations. The second set is a multiplex containing three pairs of primers for the analysis of A8344G, T8356C, G8363A, T8993G, T8993C, and G11778A. The third set of PCR is also a single PCR for the analysis of G14459A and T14484C (*see* **Note 2**).

1. Single PCR: two single PCR sets are performed. The PCR setup is similar to the one described under **Subheading 3.1.**, except that here, each PCR reaction is 25 μL, with each reagent reduced accordingly for one-quarter volume, and the primers used for one set are mtF3130 and mtR3758, for the other set are mtF14437 and mtR15185 (*see* **Table 1**). The mutations to be analyzed are A3243G, T3271C, and G3460A in one set and G14459A and T14484C in the other.
2. Multiplex PCR *(4)*: the multiplex PCR contains three pairs of primers. The PCR setup is similar to that described under **Subheading 3.1.**, except that here, the PCR reaction is 50 μL, and the primers used are mtF8278 and mtR8475 for A8344G, T8356C, and G8363A; mtF8768 and mtR9199 for T8993G and T8993C; and mtF11688 and mtR12360 for G11778A (*see* **Table 1**) (*see* **Note 3**).
3. PCR conditions for the multiplex PCR:
    a. 94°C, 2 min.
    b. 94°C, 1 min; 55°C, 1 min; 72°C, 2 min, for 30 cycles.

Table 3
Reaction Setup for Synthetic Positive Controls

Components in a total of 100 μL reaction	Volume used	Final concentration
Genomic DNA (50 ng/μL)	2 μL	1 ng/μL
10X polymerase chain reaction buffer II (ABI)	10 μL	1X
dNTP (8 mM)	2.5 μL	0.2 mM each
MgCl$_2$ (25 mM)	12 μL	3.0 mM
mtFT3271C and mtR3758 (10 μM each)	2 μL each	0.2 μM each
AmpliTaq polymerase (ABI) (5 U/μL)	0.25 μL	1.25 U total
Sterile H$_2$O	69.5 μL	

  c. 72°C, 5 min.
  d. 4°C soak.
 4. PCR conditions for the single PCRs:
  a. 94°C, 2 min.
  b. 94°C, 30 s; 55°C, 1 min; 72°C, 1 min, for 30 cycles.
  c. 72°C, 5 min.
  d. 4°C soak.

3.3. Checking Products of Amplification

1. 4 μL of PCR products are mixed with 2 μL of 3X loading dye and analyzed on a 2% agarose gel using 100-bp DNA markers.
2. The DNA bands are visualized with an ultraviolet (UV) light box.

3.4. Preparation of Dot Blots (9,10)

1. There are a total of 17 blots. Name each blot as shown in **Table 4**.
2. Spot 2.0 μL of PCR product (using multi-channel pipet, if available) onto Zeta-Probe GT membrane (Bio-Rad).
3. Spot 2.0-μL specific mutant controls.
4. Air-dry for at least 15 min.
5. Denature filters for 10 min in 0.4 N NaOH.
6. Rinse filters in distilled H$_2$O.
7. Neutralize filters for 10 min in 0.2 M Tris-HCl pH 7.5.
8. Bake filters for 30 min at 80°C vacuum oven.

3.5. Pre-Hybridization

1. Prewarm hybridization oven to 65°C.

Table 4
Setup of Dot Blots

Normal	Mutant
3243nl	A3243G
	T3271C
	G3460A
8344nl	A8344G
	T8356C
	G8363A
8993nl	T8993G
	T8993C
11778nl	G11778A
14459nl	G14459A
14484nl	T14484C

2. Prewarm Church pre-hybridization and hybridization buffer and sheared salmon sperm DNA at 65°C until completely dissolved.
3. Place the filters in 15-mL screw top conical tubes (label tubes same as filters).
4. Add 3 mL Church hybridization buffer and sheared salmon sperm DNA stock solution (10 mg/mL) to a final concentration of 100 µg/mL.
5. Incubate filters at 65°C for 30 min.

3.6. End Labeling of ASO Probes With T4 Polynucleotide Kinase

1. Each labeling reaction is 3.75 µL and contains 2.75 µL of master mix and 1 µL of a 2 µ*M* ASO probe. Depending on the number of probes to be labeled, adjust the total volume of master mix accordingly. For example, if there are 17 probes to be labeled, multiply the volume of each reagent by 17.
2. The master mix for one probe labeling is prepared as follows:
   a. 0.375 µL 10X T4 Kinase Buffer.
   b. 0.375 µL T4 Polynucleotide Kinase enzyme (10,000 U/mL from New England Biolabs).
   c. 0.75 µL γ-P^{32} ATP (4500 Ci/mmol, 10 µCi/µL, total amount is 2.2 *p*moles) (ICN).
   d. 1.25 µL sterile H$_2$O.
3. Label the 0.5-mL microcentrifuge tubes according to the name of the blot as listed under **Subheading 3.4.**
4. Aliquot 2.75 µL of master mix to each tube.
5. Dilute 10 µ*M* of ASO solution to 2 µ*M*.
6. Add 1 µL of 2 µ*M* ASO to specified tube, mix well by gently tapping the tube.
7. Incubate the tubes at 37°C for 40 min and heat to 65°C for 10 min to inactivate the enzyme.

8. Add 33.75 μL of sterile H$_2$O to each tube to dilute the labeling reaction to a final volume of 37.5 μL.

9. Store at –20°C freezer. The probes are good for 30 d.

3.7. Hybridization (4,10)

1. Use standard precautions against radioactive materials, solutions, and waste.

2. Remove the prehybridization tubes from the hybridization oven and place them in the order as listed next in the Membrane column.

3. In a separate rack, thaw 10 μ*M* cold ASO, and place them in the order as to be added (cold probe column in **Table 5**) to each pre-hybridization tube; for example, the A3243G mutant cold ASO is to be added to the tube and membrane labeled 3243nl (*see* **Note 4**).

4. Check the order of the hybridization tubes (same as the prehybridization tubes) and the ASO tubes, make sure they are in correct order and position before the addition of hot labeled and cold ASO.

5. Add 20 μL of appropriate cold ASO (stock 10 μ*M*) to each tube. For example, if the normal ASO is used as hot probe, the mutant ASO should be used as cold to reduce nonspecific hybridization (*see* **Note 4**).

6. Put the 10 μ*M* cold ASO in freezer.

7. Thaw the hot labeled ASO probes. Place them in the same order as the order of the name in the membrane and the hybridization tube to be added (hot ASO probe column in **Table 5**).

8. Add 10 μL of radioactive-labeled hot ASO to each corresponding tube, making sure that the correct ASO is added to the correct hybridization tube. When adding the hot probe, add the probe to the hybridization solution and shake the tube gently to mix well. Do not drop the hot probe onto the membrane directly. It will cause high background as a result of local high concentration.

9. Hybridize blots at 65°C for 30 min then turn temperature down to 34°C. Let hybridize until the temperature reaches 34°C and continue at 34°C for at least 3 h.

3.8. X-Ray Autoradiography

1. Decant the hybridization buffer to radioactive liquid waste container.

2. Rinse filters twice in 5X SSC at room temperature (*see* **Note 5**).

3. Wash filters twice in 2X SSC for 30 min at 34°C.

4. Use a Geiger counter to check for background on the filters. Wash the filters that have high background in 2X SSC for additional 30 min at 38°C.

5. Recheck for background and, if necessary, continue to wash at higher temperature by 2–4°C increment.

6. Expose filters to X-ray film.

7. Label the film according to the order of the dot blot membranes.

3.9. Results and Data Analysis

Normal mtDNA hybridizes to the normal probes only. Samples with homoplasmic mutations hybridize with mutant probes only. Heteroplasmic mutations hybridize with both normal and mutant probes. **Figure 1** depicts an example of results.

Table 5
Preparation of Hybridization

Membrane	Hot ASO	Cold ASO
3243nl	A3243	A3243G
A3243G	A3243G	A3243
T3271C	T3271C	T3271
G3460A	G3460A	G3460
8344nl	A8344	A8344G
A8344G	A8344G	A8344
T8356C	T8356C	T8356
G8363A	G8363A	G8363
8993nl	T8993	T8993C, T8993G (10 μL each)
T8993C	T8993C	T8993, T8993G (10 μL each)
T8993G	T8993G	T8993, T8993C (10 μL each)
11778nl	G11778	G11778A
G11778A	G11778A	G11778
14459nl	G14459	G14459A
G14459A	G14459A	G14459
14484nl	T14484	T14484C
T14484C	T14484C	T14484

ASO, allele-specific oligonucleotide.

4. Notes

1. The ASO probe is usually designed as a 19-mer with the point mutation located at the middle of the ASO. If the nucleotide is a G in the mutant sequence, an antisense ASO is made to reduce the nonspecific basepairing. For example, the antisense probes are used for A3243G and A8344G. In the case of T8993G and T8993C, because both G and C mutants are to be analyzed, the sense sequence of the mutation is used.

2. It is possible to run one single multiplex PCR reaction *(4)* containing all five pairs of primers, but the background may be higher. In order to detect a low percentage of mutant heteroplasmy, the background should be minimized.

3. Conditions for multiplex PCR should always be optimized before ASO analysis is carried out. Try to design the primers such that the PCR products can be easily separated and identified on agarose gel.

4. The reason for using a cold ASO counter part is to reduce the background. Therefore, if the background is high, the amount of the cold ASO counter part can be increased. The optimal concentration of the cold ASO to be used should be tested out for each probe.

5. The washing temperature depends on the melting temperature (T_m) of the ASO. A temperature of 34°C for washing works for most of the ASOs. Occasionally,

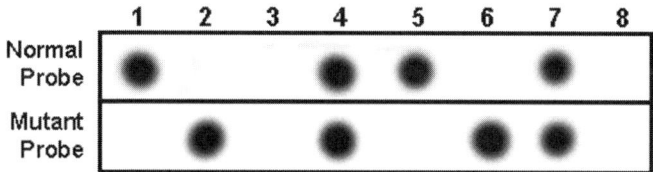

Fig. 1. Example of allele-specific oligonucleotide dot blot hybridization results.
Lane 1,: wild-type control; lane 2, homoplasmic mutant control; lane 3, no DNA template control; lane 4, heteroplasmic mutant control; lane 5, sample with wild-type mitochondrial DNA; lane 6, sample with homoplasmic mutation; lane 7, sample with heteroplasmic mutation; lane 8, sample DNA failed to amplify.

increasing the temperature may be necessary. If this is the case, gradually increase the wash temperature by 2–4°C each time, until the background is low.

References

1. Shoffner, J. and Wallace, D. (1995) *Oxidative phosphorylation diseases*, in. McGraw-Hill, New York: pp. 1535–629.
2. Smeitink, J., van den Heuvel, L., and DiMauro, S. (2001) The genetics and pathology of oxidative phosphorylation. Nat. *Rev. Genet.* **2,** 342–352.
3. Wallace, D. C. (1992) Diseases of the mitochondrial DNA. *Annu. Rev. Biochem.* **61,** 1175–1212.
4. Wong, L. J. and Senadheera, D. (1997) Direct detection of multiple point mutations in mitochondrial DNA. *Clin. Chem.* **43,** 1857–1861.
5. Wong, L. J. and Lam, C. W. (1997) Alternative, noninvasive tissues for quantitative screening of mutant mitochondrial DNA. *Clin. Chem.* **43,** 1241–1243.
6. Lahiri, D. K. and Nurnberger, J. I., Jr. (1991) A rapid non-enzymatic method for the preparation of HMW DNA from blood for RFLP studies. *Nucleic Acids Res.* **19,** 5444.
7. Liang, M. H. and Wong, L. J. (1998) Yield of mtDNA mutation analysis in 2,000 patients. *Am. J. Med. Genet.* **77,** 395–400.
8. Liang, M. H., Johnson, D. R., and Wong, L. J. (1998) Preparation and validation of PCR-generated positive controls for diagnostic dot blotting. *Clin. Chem.* **44,** 1578–1579.
9. DeMarchi, J. M., Richards, C. S., Fenwick, R. G., Pace, R., and Beaudet, A. L. (1994) A robotics-assisted procedure for large scale cystic fibrosis mutation analysis. *Hum. Mutat.* **4,** 281–290.
10. DeMarchi, J. M., Beaudet, A. L., Caskey, C. T., and Richards, C. S. (1994) Experience of an academic reference laboratory using automation for analysis of cystic fibrosis mutations. *Arch. Pathol. Lab. Med.* **118,** 26–32.

13

Novel Applications of Polymerase Chain Reaction to Urinary Nucleic Acid Analysis

Anatoly V. Lichtenstein, Hovsep S. Melkonyan, L. David Tomei, and Samuil R. Umansky

Summary

DNA fragments from cells that have died throughout the body not only appear in the bloodstream but also cross the kidney barrier into the urine. The relatively low molecular weight (150–200 bp) of this Transrenal DNA should be considered when deciding on methods of isolation and analysis. In particular, if polymerase chain reaction (PCR) is used for amplification and detection of specific sequences, then the reduction of amplicon size will significantly enhance sensitivity. Detection of DNA mutations is also made more difficult by the presence of a large excess of a wild-type allele. Using K-*RAS* mutations as an example, two ways around this problem—enriched PCR and stencil-aided mutation analysis—are described, based on selective pre-PCR elimination of wild-type sequences.

Key Words: Transrenal DNA (Tr-DNA); urine; cancer; DNA markers; K-*RAS* oncogene; mutations.

1. Introduction

Each day, 10^{11}–10^{12} cells in the human body die as a result of physiological and pathological processes. Cell death is normally accompanied by DNA degradation, and it is currently well established that a portion of resulting DNA fragments appears in the bloodstream as so-called cell free circulating DNA (for review, *see* **refs. *1–7***). It has also been found that circulating DNA fragments from blood cross the kidney barrier into the urine (*8*). This Transrenal DNA (Tr-DNA) contains sequences from cells that have died throughout the body, including tumor cells (*8,9*), developing fetus cells (*8,10*), or cells from transplanted organs (*8,11,12*). The range of potential diagnostic applications of circulating and Tr-DNA is very broad and includes tumor diagnostics and

From: *Methods in Molecular Biology, vol. 336: Clinical Applications of PCR*
Edited by: Y. M. D. Lo, R. W. K. Chiu, and K. C. A. Chan © Humana Press Inc., Totowa, NJ

monitoring, prenatal genetic testing, transplantation monitoring, analysis of age-related accumulation of DNA mutations, and so on.

There are several features of Tr-DNA that should be taken into account when developing diagnostic applications based on Tr-DNA analysis. First, the relatively low concentration of Tr-DNA in the urine must be considered *(4,8)*. In many cases, sensitive methods such as multi-round polymerase chain reaction (PCR) should be used for detection of specific sequences. Second, Tr-DNA fragments are relatively short, having an average molecular weight of approx 150–200 bp *(8)*, which dictates use of amplicons that are as short as possible. Third, if a target is a specific DNA mutation, one should consider whether a great excess of wild-type alleles is present. In the presence of a large excess of wild-type alleles, a significantly lower sensitivity of mutant DNA detection will likely result because of competition of wild-type and mutant sequences for primers. Furthermore, the presence of a great excess of wild-type sequences leads to the increased frequency of false-positive results owing to *Taq* polymerase errors *(13,14)*.

Here, we describe our experiences and provide recommendations in several specific areas of Tr-DNA analysis including preparation of urine samples designed to protect DNA from additional degradation, methods of Tr-DNA isolation, and detection of a mutant DNA sequence. As a model application, we used the detection of mutant K-*RAS* in the urine of cancer patients. Mutations in codon 12 or 13 of K-*RAS* occur in approximately half of all colorectal cancers *(15,16)*, 90% of pancreatic cancer *(16)*, and approx 30–40% of lung adenocarcinoma *(17)*. In these studies, two approaches were used for detection of mutant K-*RAS* in a great excess of wild-type allele: (1) stencil-aided mutation analysis (SAMA) as a pre-PCR procedure aimed at reducing the concentration of wild-type alleles directly in the DNA sample *(18)*; and (2) enriched PCR, which includes selective restriction-enzyme digestion of wild-type sequences after several initial PCR cycles and subsequent amplification of fragments containing the mutant codon *(19)*.

2. Materials

1. Restriction enzymes and buffers.
2. Oligonucleotide primers.
3. Reagents for PCR:
   a. *Taq* polymerase.
   b. Mixture of deoxynucleotide triphosphates.
   c. 10X PCR Buffer.
4. Polyacrylamide gel electrophoresis equipment.
5. Tris-borate-ethylenediamine tetraacetic acid (EDTA) buffer for electrophoresis (5X TBE).

6. 0.5 *M* EDTA, 0.5 *M* Tris-HCl, pH 8.5.
7. Wizard Resin Suspension and minicolumns (Promega, Madison, WI).
8. 5-mL Disposable syringes.
9. Disposable tubes (1.5 mL and 50 mL).
10. 6 *M* Guanidine isothiocyanate (GITC): 30 m*M* Tris-HCl, pH 7.0 (6 *M* GITC).
11. Column wash solution (CWS): 150 m*M* NaCl, 1 m*M* EDTA, 10 m*M* Tris-HCl, pH 7.5, 60% ethanol.
12. Nuclease-free water or TE (10 m*M* Tris-HCl, pH 8.0, 1 m*M* EDTA) for DNA elution.

3. Methods

3.1. Urine Collection

All urine specimen preparation steps for Tr-DNA analysis are carried out at room temperature unless otherwise stated. Routinely, 50–60 mL of urine were collected for each specimen (*see* **Note 1**). Immediately, but not later than 30 min after collection, 0.5 *M* EDTA-0.5 *M* Tris-HCl, pH 8.5 was added to reach a final concentration of 10 m*M* (*see* **Note 2**). Urine specimens were stored in 5-mL aliquots at –20°C or for long-term storage at –80°C. In some instances, cell-depleted urine was prepared before freezing by centrifugation for 10 min at 800–1000*g* (*see* **Note 3**).

3.2. DNA Isolation

1. All steps of DNA purification are carried out at room temperature. 20 mL of 6 *M* GITC was added to 10 mL of urine in a 50-cc tube and mixed vigorously. Usually, this step will solubilize particulate material that may be present in normal urine specimens.
2. For DNA adsorption, 0.2–0.25 mL Wizard Resin Suspension was then added and the tubes placed on a rotating shaker for at least 2 h (*see* **Note 4**). This incubation step can be extended without significantly changing in DNA yield—we routinely carried out this step overnight.
3. The resin was pelletted by centrifugation at room temperature for 10 min at 2000*g*, after which the supernatant was carefully aspirated and discarded. The resin pellet was resuspended in 1.0 mL of 4 *M* GITC, and transferred into a 1.5-mL microcentrifuge tube. This was then centrifuged in a benchtop microcentrifuge at the top speed for 30–60 s. Again, the supernatant was removed and discarded.
4. The pellet was resuspended in 1.0 mL of washing buffer and transferred in suspension to a Promega Wizard minicolumn according to the recommendations of the manufacturer. The column was then washed with an additional 2 mL of washing buffer. To remove residual washing buffer from the columns, they were centrifuged for 2–3 min in a microcentrifuge at the maximum speed.
5. The DNA was then eluted from each column with 50 µL of water or TE. In order to facilitate elution, each column–tube assembly can be incubated for 3–5 min in

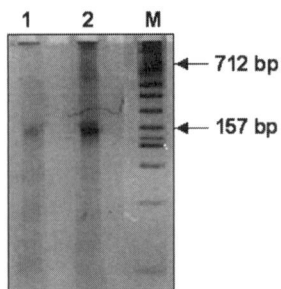

Fig. 1. Polyacrylamide gel electrophoresis of cell-free DNA isolated from human urine.

Urine samples were centrifuged for 10 min at 800*g*. DNA was isolated from the supernatants and subjected to agarose gel electrophoresis. Lanes 1 and 2, DNA from the urine of two volunteers; lane 3, molecular weight markers.

a heater set at 56°C. The yield of DNA purified according to this protocol varied significantly among different specimens providing a final range of 5 to 200 ng DNA/mL of urine. **Figure 1** illustrates that DNA isolated from cell-depleted urine specimens consists of two major fractions: a low-molecular-weight DNA whose size is close to that of mononucleosomal DNA, and high-molecular-weight DNA that most likely originates from residual cells that were not eliminated by mild centrifugation of urine samples.

3.3. Determination of K-RAS Mutations

3.3.1. Enriched PCR

K-*RAS* mutations were detected by restriction-enriched PCR (RE-PCR), a method developed for detection of a mutant codon 12 K-*RAS* allele in the presence of as many as 10^4 copies of the wild-type codon 12 allele *(19)*. This method employs a two-stage PCR. During the first stage, both mutant and wild-type alleles are amplified and then wild-type sequences are selectively digested with restriction endonuclease at an artificially created site. In the second stage, undigested PCR products enriched in mutant codon 12 sequences are amplified. This technique was originally developed for the analysis of K-*RAS* mutation in DNA extracted from tumor cells. Therefore, the amplicon size of 157 bp proposed by the authors (**Fig. 2**) was not well suited for analysis of Tr-DNA, because these fragments themselves are about 150 bp in size (*see* **Note 5**). To adjust RE-PCR for analysis of Tr-DNA, we modified the existing protocol by designing a new reverse primer targeting a shorter (87 bp) DNA fragment (*see* **Figs. 2** and **3** and **refs.** *20,21*). For comparison, PCR were carried out with two sets of primers in parallel: K-*RAS*-F (5'-actgaatataaacttgtggtagtt ggacct-3') paired with K-*RAS*-RL (5'-tcaaagaatggtcctgcacc-3') or with K-*RAS*-RS

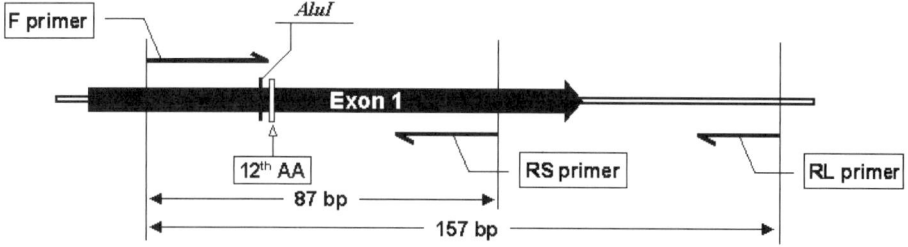

Fig. 2. Schematic view of the K-*RAS* fragment amplified by forward (F) and reverse primers producing 87-bp (RS) and 157-bp (RL) DNA fragments, respectively.

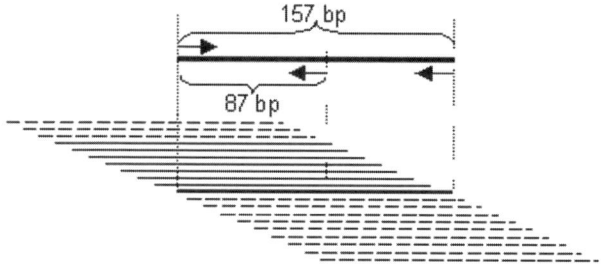

Fig. 3. Dependence of polymerase chain reaction sensitivity on amplicon size.

Different lines illustrate short DNA fragments (the 157-bp fragment is given as an example) generated by random cleavage of K-*RAS* in the area of codon 12. Bold solid line represents the only fragment that is amplified by primers designed for the 157-bp amplicon. Thin solid lines represent subset of DNA fragments amplified by the pair of primers targeting the 87-bp amplicon. Dashed lines are DNA fragments that are not amplified by either set of primers.

(5'-gtccacaaaatgattctgaattagc-3') amplifying 157- and 87-bp DNA fragments, respectively. The first primer, which is immediately upstream of codon 12, was modified at nucleotide 28 (G to C) to create a restriction enzyme site specific for B*st* NI in wild-type K-*RAS* DNA. Any mutations of the first or second nucleotide of codon 12 destroy this restriction site that can be seen by mapping of the DNA fragment by B*st* NI cleavage sites. For the second stage, reverse primers are also modified to introduce additional B*st* NI cleavage site for control of the completeness of restriction: K-*RAS*-RL' (5'-tcaaagaatggtcctggacc-3') and K-*RAS*-S' (5'-gtccacaaaatgatcctggattagc-3'). The analysis of K-*RAS* mutations was carried out as follows.

1. One-fiftieth of DNA purified from 10 mL urine was subjected to the first step of PCR. 25-µL reaction mixtures were cycled 20 times at 94°C for 30 s, 56°C for 30 s, and 72°C for 30 s.

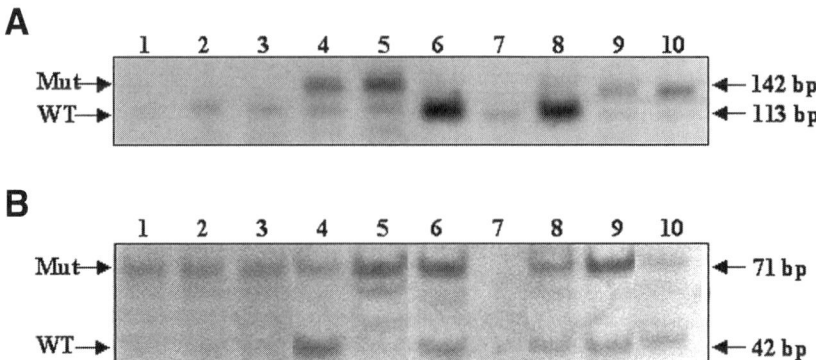

Fig. 4. Detection of K-*RAS* mutation by enriched polymerase chain reaction in Tr-DNA of patients with pancreatic cancer. (**A,B**) Amplification of 157 and 87-bp fragments, respectively.

2. An aliquot of 10 µL adjusted to 1X B*st* NI reaction buffer in 25 µL was digested with 10 U of B*st* NI at 60°C for 90 min.
3. 1 µL of the digested PCR mixture was transferred to a new tube and a fresh 25-µL reaction mixture was set up for the second amplification stage.
4. Probes were cycled 35 times with the same thermal profile as the first PCR, but with different reverse primers (discussed previously).
5. Finally, diagnostic B*st* NI restriction was performed using 10 µL of the second-step PCR product.
6. The digestion products were separated by electrophoresis through 7 or 12% poly-acrylamide gels (**Fig. 4**). Obviously, the sensitivity of the reaction with smaller amplicon is much higher. K-*RAS* mutation, undetectable in samples 1–3, 6, and 8 with the 157-bp amplicon, becomes evident with the shorter amplicon.

3.3.2. Pre-PCR Digestion of Wild-Type K-RAS Alleles (SAMA)

Selective digestion of wild-type DNA was achieved by a combination of high-stringency molecular hybridization of wild-type complementary stencil oligonucleotide to Tr-DNA with simultaneous restriction endonuclease cleavage of hybridized duplexes. Conditions of annealing and nuclease digestion prevent formation of double-stranded structure with the mutant sequence and leads to preferential cleavage of wild-type allele (**Fig. 5**). To apply this technique for testing of codon 12 mutation of K-*RAS* gene, we took advantage of the presence of *Alu*I digestion site in close vicinity to the site of interest. Treatment of DNA samples was carried out in a two-step reaction.

1. In the first step, the minus strand of wild-type K-*RAS* was cleaved by *Alu*I using a plus stencil oligonucleotide. Plus-stencil is a 16-bp oligonucleotide (5'-AGTTGGAGCTGGTGGC-3') spanning the wild-type immediately upstream

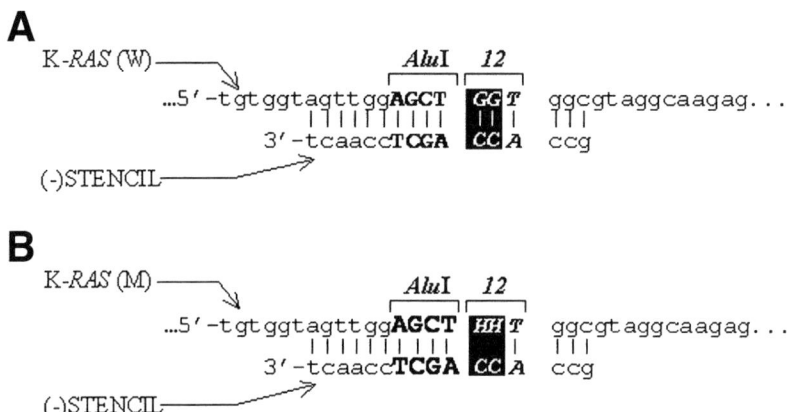

Fig. 5. Sequence alignment of stencil oligonucleotide with wild-type and mutated K-*RAS* DNA.

*Alu*I restriction site near codon 12 is shown in bold. H designates any nucleotide except G.

from codon 12. Six nucleotides flanking both sides of the *Alu*I recognition site were required for the enzyme to bind and efficiently cleave the target. The DNA sample (1–1000 ng) in 60 µL of *Alu*I restriction endonuclease buffer was mixed with 20 pmol of plus-stencil, heated at 95°C for 10 min, annealed at 47°C for 10 min and digested by *Alu*I (5 U) at 42.5°C for 2 h.

2. At the second step, the plus strand of wild-type K-*RAS* was cleaved similarly using the minus-stencil (5'-GCCACCAGCTCCAACT-3'), the latter being complementary to the plus-stencil, specified previously. The twofold amount (i.e., 40 pmol) of minus-stencil in *Alu*I buffer was added to the DNA sample heated at 95°C for 10 min (the excess of the minus-stencil is needed to ensure the complete hybridization with template DNA in the presence of the complementary plus-stencil). Denaturation, annealing, and digestion steps were repeated as described.

3. Finally, the restriction enzyme was inactivated by heating at 95°C for 10 min, DNA was precipitated with ethanol, washed with 70% ethanol, dissolved in water, and subjected to PCR.

Figure 6 presents the results of experiments with artificial mixtures of wild-type and mutant K-*RAS*. After SAMA, the band specific for mutant K-*RAS* was not reduced, whereas the band of wild-type K-*RAS* disappeared. At the highest wild-type/mutant sequences ratio (1000/1) the competition of mutant K-*RAS* with the large excess of wild-type allele in the control sample considerably weakened the mutant signal. However, the stencil-aided pre-PCR digestion of the wild-type sequences made the mutant K-*RAS* band much stronger. Application of SAMA to three urine samples with undetectable by RE-PCR levels of mutant K-*RAS* revealed mutation in three urine specimens (**Fig. 7**).

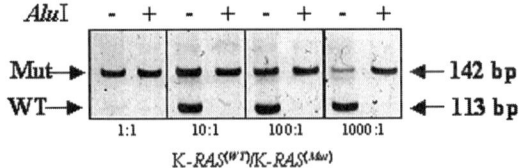

Fig. 6. Stencil-aided mutation analysis (SAMA) of mutant K-*RAS* DNA in the presence of different amounts of wild-type allele. One nanogram of DNA from SW480 cell line carrying two mutated alleles at codon 12 of the K-*RAS* gene was mixed with different amounts of DNA from peripheral lymphocytes of healthy donors and subjected to SAMA following by enriched polymerase chain reaction (157-bp amplicon).

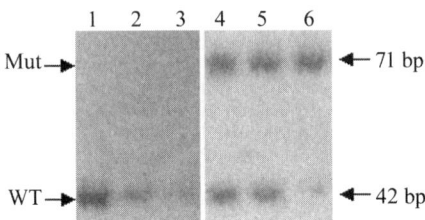

Fig. 7. Detection of K-*RAS* mutation in Tr-DNA from three patients with pancreatic cancer.

Lanes 1–3, enriched polymerase chain reaction (PCR); lanes 4–6, stencil-aided mutation analysis (SAMA) plus enriched PCR.

4. Notes

1. We have not observed a significant difference between DNA isolated from urine specimens obtained at different time points during the day. However, we prefer not to use the first morning urine specimens, because additional DNA degradation may have occurred during the hours of sleep.
2. It is highly desirable to inhibit nucleases that have been found in some urine samples. EDTA is added to inhibit DNase-I type nucleases, and a relatively high pH inhibits acidic nucleases.
3. If there is not the specific purpose of analyzing cell free DNA, we would normally isolate DNA from unfractionated urine specimens because a portion of cell free DNA has a tendency to co-sediment with cells during centrifugation. Although this effect cannot be entirely prevented, it can be minimized by centrifugation at room temperature and washing the sediment.
4. To avoid a very significant loss of Tr-DNA, commercial kits designed for isolation of high molecular weight genomic DNA should not be used for Tr-DNA purification.
5. **Figure 3** illustrates the advantages of a shorter amplicon size when PCR is performed with small DNA templates. Most of the long amplicons are out of

frame determined by the respective primers. It is anticipated that DNA fragments are generated by random cleavage of nuclear DNA.

References

1. Taback, B. and Hoon, D. S. (2004) Circulating nucleic acids in plasma and serum: past, prsent and future. *Curr. Opin. Mol. Ther.* **6,** 273–278.
2. Garcia-Olmo, D. C., Ruiz-Piqueras, R., and Garcia-Olmo, D. (2004) Circulating nucleic acids in plasma and serum (CNAPS) and its relation to stem cells and cancer metastasis: state of the issue. Quantification of circulating DNA in the plasma and serum of cancer patients. *Histol. Histopathol.* **19,** 575–583.
3. Wong, B. C. and Lo, Y. M. D. (2003) Cell-free DNA and RNA in plasma as new tools for molecular diagnostics. *Expert Rev. Mol. Diagn.* **3,** 785–797.
4. Lichtenstein, A. V., Melkonyan, H. S., Tomei, L. D., and Umansky, S. R. (2001) Circulating nucleic acids and apoptosis. *Ann. N. Y. Acad. Sci.* **945,** 239–249.
5. Anker, P., Mulcahy, H., and Stroun, M. (2003) Circulating nucleic acids in plasma and serum as a noninvasive investigation for cancer: time for large-scale clinical studies? *Int. J. Cancer* **103,** 149–152.
6. Goessl, C. (2003) Diagnostic potential of circulating nucleic acids for oncology. *Expert Rev. Mol. Diagn.* **3,** 431–442.
7. Chan, K. C. A., Chiu, R. W. K., and Lo, Y. M. D. (2003) Cell-free nucleic acids in plasma, serum and urine: a new tool in molecular diagnosis. *Ann. Clin. Biochem.* **40(Pt 2),** 122–130.
8. Botezatu, I., Serdyuk, O., Potapova, G., et al. (2000) Genetic analysis of DNA excreted in urine: a new approach for detecting specific genomic DNA sequences from cells dying in an organism. *Clin. Chem.* **46,** 1078–1084.
9. Utting, M., Werner, W., Dahse, R., Schubert, J., and Junker K. (2002) Microsatellite analysis of free tumor DNA in urine, serum, and plasma of patients: a minimally invasive method for the detection of bladder cancer. *Clin. Cancer Res.* **8,** 35–40.
10. Al-Yatama, M. K., Mustafa, A. S., Ali, S., Abraham, S., Khan, Z., and Khaja, N. (2001) Detection of Y chromosome-specific DNA in the plasma and urine of pregnant women using nested polymerase chain reaction. *Prenat. Diagn.* **21,** 399–402.
11. Zhang, J., Tong, K. L., Li, P. K., et al. (1999) Presence of donor- and recipient-derived DNA in cell-free urine samples of renal transplantation recipients: urinary DNA chimerism. *Clin. Chem.* **45,** 1741–1746.
12. Zhang, Z., Ohkohchi, N., Sakurada, M., et al. (2002) Analysis of urinary donor-derived DNA in renal transplant recipients with acute rejection. *Clin. Transplant.* **16,** 45–50.
13. Smith, J. and Modrich, P. (1997) Removal of polymerase-produced mutant sequences from PCR products. *Proc. Natl. Acad. Sci. USA* **94,** 6847–6850.
14. Jacobs, G., Tscholl, E., Sek, A., Pfreundschuh, M., Daus, H., and Trumper, L. (1999) Enrichment polymerase chain reaction for the detection of Ki-ras mutations: relevance of Taq polymerase error rate, initial DNA copy number, and reaction conditions on the emergence of false-positive mutant bands. *J. Cancer Res. Clin. Oncol.* **125,** 395–401.

15. Tobi, M., Luo, F. C., and Ronai, Z. (1994) Detection of K-ras mutation in colonic effluent samples from patients without evidence of colorectal carcinoma. *J. Natl. Cancer Inst.* **86,** 1007–1010.

16. Bos, J. L. (1989) Ras oncogenes in human cancer: a review. *Cancer Res.* **49,** 4682–4689.

17. Ahrendt, S. A., Chow, J. T., Xu, L. H., et al. (1999) Molecular detection of tumor cells in bronchoalveolar lavage fluid from patients with early stage lung cancer. *J. Natl. Cancer Inst.* **91,** 332–339.

18. Lichtenstein, A. V., Serdjuk, O. I., Sukhova, T. I., Melkonyan, H. S., and Umansky, S. R. (2001) Selective 'stencil'-aided pre-PCR cleavage of wild-type sequences as a novel approach to detection of mutant K-RAS. *Nucleic Acids Res.* **29,** E90.

19. Kahn, S. M., Jiang, W., Culbertson, T. A., et al. (1991) Rapid and sensitive nonradioactive detection of mutant K-ras genes via 'enriched' PCR amplification. *Oncogene* **6,** 1079–1083.

20. Su, Y.-H., Wang, M., Brenner, D. E., et al. (2004) Human urine contains small, 150-250 nucleotide sized, soluble DNA derived from the circulation and may be useful in the detection of colorectal cancer. *J. Mol. Diagn.* **6,** 101–107.

21. Su, Y.-H., Wang, M., Block, T. M., et al. (2004) Transrenal DNA as a diagnostic tool: important technical notes. *Ann. New York Acad. Sci.* **1022,** 81–89.

14

Detection and Quantitation of Circulating *Plasmodium falciparum* DNA by Polymerase Chain Reaction

Shira Gal and James S. Wainscoat

Summary

This chapter describes the application of polymerase chain reaction (PCR) for the detection and quantitation of *Plasmodium falciparum* DNA in the plasma of malaria-infected individuals. The procedure includes the following protocols: plasma sample preparation, DNA extraction, detection of *P. falciparum* DNA in the plasma by nested PCR, and quantitation of *P. falciparum* DNA in the plasma by real-time PCR technology.

Key Words: *Plasmodium falciparum*; malaria; plasma; PCR; real-time PCR.

1. Introduction

The standard method for the diagnosis of malaria is the preparation and microscope examination of stained blood smears *(1)*. However, in recent years, polymerase chain reaction (PCR) has been introduced for the detection of DNA from the *Plasmodium* parasite. The major advantage of sensitive PCR techniques is the ability to detect a low level of parasitemia. Detection of as few as one parasite per microliter of blood have been reported *(2–4)*. Other applications of PCR in malaria studies include the detection of mixed infections *(3,5)* and the identification of different genotypes and genetic markers of virulence *(6)*. The introduction of real-time PCR technology has enabled a quantitative method for the measurement of parasite burden. Real-time PCR was used for several purposes: monitoring the levels of *P. falciparum* in the blood of malaria-infected individuals *(7)*, the quantitation of malaria liver-stage parasites in mice *(8,9)*, and the detection of malaria parasites in blood of infected people *(10)*. The following protocol describes the application of PCR and real-time PCR for the detection and quantitation of circulating *P. falciparum* DNA in the plasma of malaria patients.

From: *Methods in Molecular Biology, vol. 336: Clinical Applications of PCR*
Edited by: Y. M. D. Lo, R. W. K. Chiu, and K. C. A. Chan © Humana Press Inc., Totowa, NJ

2. Materials

2.1. Equipment

1. Centrifuge.
2. Micro-centrifuge.
3. Vortex.
4. Water bath.
5. PCR machine.
6. Agarose gel electrophoresis equipment.
7. Sequence Detection System (real-time PCR).
8. Pipets for PCR.
9. Filter tips.

2.2. Reagents

1. QIAamp DNA Blood Mini Kit (Qiagen).
2. PCR kit.
3. Primers and probes.
4. TaqMan Universal Master Mix (Applied Biosystems).
5. MicroAmp Optical 96-well Reaction Plate (Applied Biosystems).
6. MicroAmp Adhesive Covers or Optical Caps (Applied Biosystems).
7. Nuclease-free water.

3. Methods

The methods described in **Subheadings 3.1.–3.4.** outline (1) blood collection and plasma separation, (2) extraction of DNA from plasma, (3) detection of *P. falciparum* parasite using a nested PCR assay, and (4) quantitation of *P. falciparum* DNA in plasma by real-time PCR.

3.1. Blood Collection and Plasma Separation

1. Collect blood into ethylenediamine tetraacetic acid (EDTA) tubes. Keep blood samples at room temperature and process the blood as soon as possible.
2. Centrifuge samples at 1000*g* for 10 min at room temperature.
3. Transfer the plasma layer into a clean, labeled, 15-mL falcon tube. Take extra care to not disturb the buffy coat layer.
4. Spin the plasma again (as in **step 2**).
5. Remove the plasma without disturbing the pellet and transfer it into clean, labeled, DNase-free tubes and store at –70°C.

3.2. Extraction of DNA From Plasma

DNA is extracted from plasma using the QIAamp Blood Mini Kit (Qiagen). The protocol is described as follows, but for more details and troubleshooting, please refer to the kit handbook.

3.2.1. Reagent Preparation

1. Add absolute ethanol to buffers AW1 and AW2 as directed on the bottles.
2. Dissolve the protease with the protease solvent and divide the ready protease into aliquots. Store ready-to-use protease at –20°C until first use. Thaw protease before use and adjust it to room temperature. After having been thawed once, the protease should be kept at 4°C, at which it is stable for up to 2 mo.
3. Heat the water bath to 56°C.

3.2.2. Samples and Labeling

1. For each sample label one 15-mL tube, one 1.5-mL microfuge tube, and one QIAamp spin column.
2. Thaw the protease and adjust it to room temperature.
3. Thaw samples and adjust them to room temperature.
4. Sample volume (X) can range between 200 and 1000 μL. It is best to use the same volume for all of the samples being analyzed. When using 400 μL or less of sample, the procedure should be carried out using 1.5-mL microfuge tubes, and brief spins should be performed before **steps 5** and **6**.
5. The procedure is carried out at room temperature.

3.2.3. Protocol

1. Pipet 0.1X μL of protease to each of the 15-mL tubes (or 1.5-mL tubes, if sample volume in less than 400 μL).
2. Add X μL of plasma to the matching 15-mL tube.
3. Add X μL of buffer AL and vortex for 15 s.
4. Incubate the samples in a water bath of 56°C for 10 min.
5. Add X μL of absolute ethanol to each tube, vortex for 15 s.
6. Load 630 μL of the mixture on the spin column and centrifuge for 1 min at 8000*g* at room temperature.
7. Transfer the column to a new collecting tube (supplied with the kit). If more than 200 μL of sample is processed, this step should be repeated until all of the mixture is loaded on the column. Discard the flow-through before each spin.
8. Add 500 μL of Buffer AW1 to each column and centrifuge at 8000*g* for 1 min at room temperature.
9. Transfer the columns into new collecting tubes.
10. Add 500 μL of Buffer AW2 to each column and centrifuge at maximum speed for 3 min.
11. Carefully transfer the columns to a clean, labeled, 1.5-mL Eppendorf tube, ensure that there are no traces of the ethanol in the column, and discard the flow-through to a special container for sodium azide waste.
12. Add 50 μL of Buffer AE to each column. Incubate for 5 min at room temperature.
13. Centrifuge at 8000*g* for 1 min at room temperature to elute the DNA from the columns.
14. Discard the columns and store the samples at –20°C.

3.3. Detection of P. falciparum Parasite Using a Nested PCR Assay

Described as follows is a nested PCR assay for the detection of *P. falciparum* DNA in the plasma of malaria-infected individuals *(11)*. The PCR assay was developed by Snounou et al. *(2,3)* for the amplification of the *small sub-unit rRNA (ssrRNA)* gene of *P. falciparum* DNA. The first amplification round is genus-specific, and thus will amplify all species of the *Plasmodium* parasite, whereas the second amplification round is species-specific (*see* **Notes 1** and **2**).

1. Perform PCR in a total volume of 50 µL and include the extracted plasma DNA (5 µL), 1X *Taq* polymerase buffer, 3 m*M* of $MgCl_2$, dNTP mix (500 n*M*), forward and reverse primer (50 pmol of each primer), 2.5 U of *Taq* polymerase.
2. Suggested thermal profile:

	95°C	5 min
35 cycles of:	94°C	30 s
	58°C	1 min
	72°C	1 min
	72°C	5 min

3. For the nested PCR reaction, use 2 µL of the first-round amplification and follow the same thermal profile (*see* **Note 3**).
4. Primer sequences are as follows:
   a. First amplification Forward: 5' TTA AAA TTG TTG CAG TTA AAA CG 3'.
   b. First amplification Reverse: 5' CCT GTT GTT GCC TTA AAC TTC 3'.
   c. Second amplification Forward (*P. falciparum*): 5' TTA AAC TGG TTT GGG AAA ACC AAA TAT AT T 3'.
   d. Second amplification Reverse (*P. falciparum*): 5' ACA CAA TGA ACT CAA TCA TGA CTA CCC GTC 3'.
5. An amplification reaction without DNA template should be included in each PCR experiment to exclude false-positives resulting from contamination of reagents. A malaria-positive sample should be included as a positive control. In addition to PCR amplification of the *ssrRNA* gene, all DNA samples should be subjected to PCR amplification of a human gene in order to confirm the presence of intact DNA in the samples.
6. The presence of the second round amplification product (205 bp) can be detected by ethidium bromide staining following agarose gel electrophoresis (**Fig. 1**).

3.4. Quantitation of P. falciparum DNA in Plasma by Real-Time PCR

Described as follows is the protocol for the quantitation of parasite DNA from plasma (*see* **Note 2**) using a 5700 Sequence Detection System (Applied Biosystems). The procedure can be performed with an equivalent instrument using the appropriate reagents.

3.4.1. Preparation of the Reagents

1. Standard curve: for absolute quantitation of parasite DNA, a standard curve should be constructed using serial dilutions of known concentrations of parasite

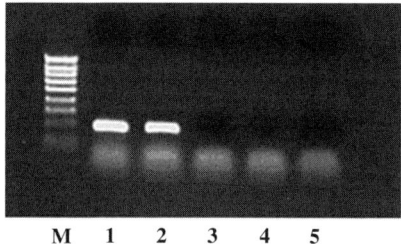

M   1   2   3   4   5

Fig. 1. Polymerase chain reaction (PCR) amplification for the *small sub-unit rRNA*
(*ssrRNA*) gene of *Plasmodium falciparum*. Amplification bands of nested PCR visualized on 1% agarose gel stained with ethidium bromide. Lanes 1 and 2, DNA extracted
from the plasma of malaria patients; lanes 3 and 4,: DNA extracted from the plasma of
healthy individuals; lane 5: no template control.

DNA. This control DNA is obtained from the extraction of DNA from erythrocyte
culture, which has been infected with the ring form of *P. falciparum*. The DNA
extracted should be quantified by a spectrophotometer (the mean of two readings
of two dilutions). From the DNA source, 10-fold serial dilutions should be made
ranging from 100,000 to 1 parasite genome equivalent per microliter (using the
conversion factor of 1 parasite genome equivalent = 0.02 pg). Make enough volume of each dilution for the experiments, aliquot, and freeze at –20°C.
2. The TaqMan assay: the assay is specific for the *P. falciparum ssrRNA* gene. Primer
and probe sequences are as follows (GenBank accession no. M19173):
a. Fal-792F: 5'-GCT TTT GAG AGG TTT TGT TAC TTT GAG-3'.
b. Fal-889R: 5'-TTC CAT GCT GTA GTA TTC AAA CAC AAT-3'.
c. Fal-820T: 5'-(FAM) CTC AAT CAT GAC TAC CCG TCT GTT ATG AAC
ACT TAA TTT T (TAMARA)-3'.
Prepare dilutions of 5 pmols/μL for the primers and probes with nuclease-free water.

3.4.2. Real-Time PCR Experiment

Analyzing the samples and standards in triplicate is recommended, and a
nontemplate control should be included in each run.

1. Open the Sequence Detection System software and create a new file for your
experiment putting the type, name, and quantity (for standards) to each of the
96 wells in use (*see* **Note 4**).
2. Thaw all reagents and put on ice.
3. Prepare a master mix from the following reagents (make 10% extra):

Nuclease-free water	5 μL per reaction
Taqman Universal Master Mix	12.5 μL per reaction
Forward primer	1.5 μL per reaction
Reverse primer	1.5 μL per reaction
Probe	0.5 μL per reaction
Total	21 μL per reaction

4. Mix the master mix by inverting and gently flicking the tube and following with a brief pulse spin.
5. Thaw samples and standards on ice and briefly spin them before use.
6. Pipet 21 μL of the mix into each of the 96 wells in use.
7. Pipet 4 μL of standard/sample/water blank in the matching well.
8. Seal the plate using the caps or an adhesive cover and spin the plate at 130*g* for 30 s.
9. Insert the plate into the Sequence Detection System and start the run. Make sure the PCR profile is appropriate. Using the Applied Biosystems Sequence Detection System, the thermal profile is usually as follows (*see* **Note 5**):

	2 min	50°C
	10 min	95°C
40 cycles of:	10 s	95°C
	60 s	60°C

Be sure to change the PCR reaction volume to 25 μL.

3.4.3. Analysis of the Real-Time PCR Results

Notice that the Sequence Detection System software may differ between machines and versions.

1. Adjust the threshold to the middle of the linear part of the curves and analyze the amplification curves for the samples and standards. Make sure the standard curve is linear with correlation >–0.990, that the sample threshold cycle (C_T) is in the range of the standard curve, and that the standard deviation (SD) between the replicates is small; otherwise, repeat the experiment (**Fig. 2**).
2. The Sequence Detection System software should calculate the Mean Quantity (the average quantity of your triplicates for each sample according to the standard curve). This value represents the genome equivalent measurable in your sample and should be used for further analysis.
3. The DNA concentration in genome equivalents per milliliter plasma is calculated using the following formula. It is crucial to use this formula when starting with different volumes of sample.

$$C = Q \times \frac{V_{DNA}}{V_{PCR}} \times \frac{1}{V_{ext}}$$

C = target concentration in plasma (genome equivalents per milliliter).
Q = target quantity (genome equivalents) calculated by the Sequence Detection System.
V_{DNA} = total volume of extraction (50 μL).
V_{PCR} = volume of DNA used per PCR reaction (4 μL).
V_{Ext} = volume of plasma extracted (x mL).

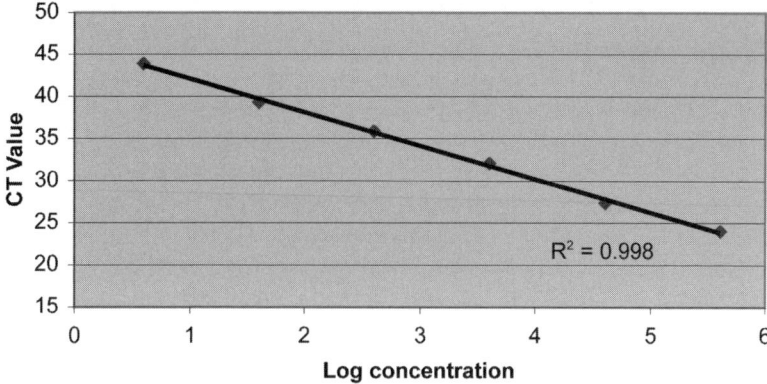

Fig. 2. An example of a standard curve for the real-time polymerase chain reaction assay of the *small sub-unit rRNA* (*ssrRNA*) gene of *Plasmodium falciparum*. Standard curves constructed by plotting the threshold cycle value (*y*-axis) against the input concentration of each dilution (*x*-axis) from 400,000 to 4 genome equivalents in logarithmic scale.

4. Notes

1. By using different second round amplification primers, it is possible to detect the other species of the *Plasmodium* parasite.
2. The nested PCR assay for the detection of *P. falciparum* DNA as well as the TaqMan assay can be performed on DNA extracted from the blood of malaria patients. DNA may be extracted using the Qiagen QIAamp blood mini kit for small volumes of blood or any other DNA extraction kit. It is sometimes recommended to extract parasite DNA from the red cells.
3. PCR conditions suggested under **Subheading 3.3.** may need adjustment—i.e., reagent concentration and thermal profile may be different when using different PCR kits and equipment.
4. Because 4 μL of sample/standard is used for each reaction, the quantities of the standards are 400,000, 40,000, 4000, 400, 40, and 4 genome equivalents.
5. Cycle number may be increased to 45 if the samples have very low parasitemia.

References

1. Makler, M. T., Palmer, C. J., and Ager, A. L. (1998) A review of practical techniques for the diagnosis of malaria. *Ann. Trop. Med. Parasitol.* **92,** 419–433.
2. Snounou, G., Viriyakosol, S., Jarra, W., Thaithong, S., and Brown, K. N. (1993) Identification of the four human malaria parasite species in field samples by the polymerase chain reaction and detection of a high prevalence of mixed infections. *Mol. Biochem. Parasitol.* **58,** 283–292.

3. Snounou, G., Viriyakosol, S., Zhu, X. P., et al. (1993) High sensitivity of detection of human malaria parasites by the use of nested polymerase chain reaction. *Mol. Biochem. Parasitol.* **61,** 315–320.

4. Tirasophon, W., Rajkulchai, P., Ponglikitmongkol, M., Wilairat, P., Boonsaeng, V., and Panyim, S. (1994) A highly sensitive, rapid, and simple polymerase chain reaction-based method to detect human malaria (Plasmodium falciparum and Plasmodium vivax) in blood samples. *Am. J. Trop. Med. Hyg.* **51,** 308–313.

5. Oliveira, D. A., Holloway, B. P., Durigon, E. L., Collins, W. E., and Lal, A. A. (1995) Polymerase chain reaction and a liquid-phase, nonisotopic hybridization for species-specific and sensitive detection of malaria infection. *Am. J. Trop. Med. Hyg.* **52,** 139–144.

6. Ntoumi, F., Contamin, H., Rogier, C., Bonnefoy, S., Trape, J. F., and Mercereau-Puijalon, O. (1995) Age-dependent carriage of multiple Plasmodium falciparum merozoite surface antigen-2 alleles in asymptomatic malaria infections. *Am. J. Trop. Med. Hyg.* **52,** 81–88.

7. Hermsen, C. C., Telgt, D. S., Linders, E. H., et al. (2001) Detection of Plasmodium falciparum malaria parasites in vivo by real-time quantitative PCR. *Mol. Biochem. Parasitol.* **118,** 247–251.

8. Bruna-Romero, O., Hafalla, J. C., Gonzalez-Aseguinolaza, G., Sano, G., Tsuji, M., and Zavala, F. (2001) Detection of malaria liver-stages in mice infected through the bite of a single Anopheles mosquito using a highly sensitive real-time PCR. *Int. J. Parasitol.* **31,** 1499–1502.

9. Witney, A. A., Doolan, D. L., Anthony, R. M., Weiss, W. R., Hoffman, S. L., and Carucci, D. J. (2001) Determining liver stage parasite burden by real time quantitative PCR as a method for evaluating pre-erythrocytic malaria vaccine efficacy. *Mol. Biochem. Parasitol.* **118,** 233–245.

10. Lee, M. A., Tan, C. H., Aw, L. T., et al. (2002) Real-time fluorescence-based PCR for detection of malaria parasites. *J. Clin. Microbiol.* **40,** 4343-4345.

11. Gal, S., Fidler, C., Turner, S., Lo, Y. M. D., Roberts, D. J., and Wainscoat, J. S. (2001) Detection of Plasmodium falciparum DNA in plasma. *Ann. N. Y. Acad. Sci.* **945,** 234–238.

15

Molecular Diagnosis of Severe Acute Respiratory Syndrome

Enders K. O. Ng and Y. M. Dennis Lo

Summary

The etiologic agent of severe acute respiratory syndrome (SARS) has been identified as a new type of coronavirus, known as SARS-coronavirus (SARS-CoV). Although the SARS epidemic has subsided, many authorities, including the World Health Organization (WHO) and the Centers for Disease Control and Prevention (CDC), have warned of the possible re-emergence of this highly infectious disease. Although antibody-based diagnosis of SARS has been demonstrated to be a reliable proof of SARS infection, it is not sensitive enough for detection during the early phase of the disease. To date, based on the publicly released full genomic sequences of SARS-CoV, various molecular detection methods based on reverse-transcription polymerase chain reaction (RT-PCR) have been developed. Although most of the assays have initially been focused on RNA extracted from nasopharyngeal aspirates, urine, and stools, several of the more recently developed assays have been based on the analysis of RNA extracted from plasma and serum. Such assays allow the more standardized quantitative expression of viral loads and are potentially useful for early SARS diagnosis. In this chapter, two real-time quantitative RT-PCR systems for the quantification of SARS-CoV RNA in serum are discussed. The two RT-PCR systems, one aimed toward the nucleocapsid region and the other toward the polymerase region of the virus genome, have a detection rate of up to 80% during the first week of illness. These quantitative systems are potentially useful for the early diagnosis of SARS and can also provide viral load information that might assist clinicians in making a prognostic evaluation of an infected individual.

Key Words: Serum RNA; SARS-CoV RNA; viral RNA extraction; RNA quantification; real-time quantitative reverse-transcription PCR.

1. Introduction

The identification of severe acute respiratory syndrome-coronavirus (SARS-CoV) as the etiologic agent of SARS has led scientists to develop rapid and sensitive diagnostic tests *(1–6)*. Although the SARS epidemic has subsided,

From: *Methods in Molecular Biology, vol. 336: Clinical Applications of PCR*
Edited by: Y. M. D. Lo, R. W. K. Chiu, and K. C. A. Chan © Humana Press Inc., Totowa, NJ

many authorities, including the World Health Organization (WHO) and the Centers for Disease Control and Prevention (CDC), have warned of the possible re-emergence of this highly infectious disease. Thus, the sensitive tests for SARS are of great public health importance. However, a number of diagnostic tests are not very sensitive during the early phase of the disease. For instance, the use of nasopharyngeal aspirates using an early generation reverse-transcription polymerase chain reaction (RT-PCR) for SARS-CoV has a sensitivity of only 32% on d 3 of the disease *(7)*. This severe limitation has restricted our ability to identify patients in a prompt manner and to institute isolation and treatment.

Based on the publicly released full genomic sequences of SARS-CoV *(8–10)*, various molecular detection methods based on RT-PCR have been developed. These PCR-based diagnostic tests are used to detect SARS-CoV RNA in patients' specimens in which viral RNA is reverse-transcribed into DNA and then different regions of the SARS-CoV genome are specifically amplified by PCR. Several RT-PCR protocols developed by members of the WHO laboratory network are available on the WHO website (http://www.who.int/csr/sars/primers/en/).

RT-PCR is mainly divided into qualitative (conventional) and quantitative approaches. Conventional RT-PCR approaches are normally qualitative in nature and require time-consuming and contamination-prone post-PCR analysis. Real-time quantitative RT-PCR has overcome many of these shortcomings and has been increasingly adopted by various laboratories for SARS diagnosis *(2,11–15)*. With suitable instrumentation, this technology allows data to be recorded and analyzed during PCR cycling. Furthermore, it runs as a closed-tube system, and postamplification manipulation can be eliminated. Thus, this methodology reduces the risk of contamination and minimizes hands-on time. The entire amplification process requires only 3 h and allows such technology to be used for high-throughput application.

During the SARS outbreak, the PCR-based testing for SARS was focused mainly on the analysis of nasopharyngeal aspirates, urine, and stools *(7,11)*. An early study reported that SARS-CoV RNA was detected in 32% of nasopharyngeal aspirates from SARS patients studied at a mean of 3.2 d after the onset of illness, and the detection rate increased to 68% at d 14 *(7)*. In the same study, SARS-CoV RNA was detected in 97% of stool samples collected at a mean of 14.2 d after symptom onset. Similarly, viral RNA was detected in 42% of urine samples collected from the SARS patients at a mean of 15.2 d after onset *(7)*. Despite the high sensitivity of stool sample testing, early detection of SARS-CoV still suffers from a lack of high sensitivity. Although most of the assays have been predominantly focused on RNA extracted from nasopharyngeal aspirates, urine, and stools, the quantitative interpretation of

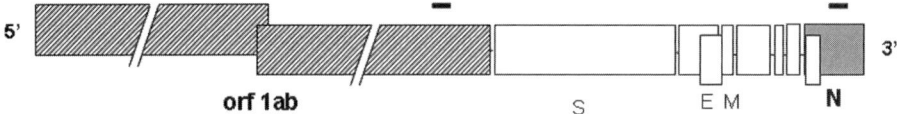

Fig. 1. The locations of the polymerase chain reaction (PCR) amplicons within the genomic organization of severe acute respiratory syndrome (SARS)-coronavirus (CoV). The black bars represent the two PCR amplicons located within the genomic structure of SARS-CoV. The size of the open reading frames (ORFs) is drawn to scale, except for orf1ab. The size of the genome is 29.3 kb. *Shaded boxes* represent ORFs encoding the viral polymerase, whereas the *filled box* (N) represents the nucleocapsid region. S represents spike protein; E represents envelope protein; M represents membrane protein.

these data can be difficult because of the inability to standardize such data as a result of numerous factors, such as sampling technique for nasopharyngeal aspirates, urine volume, variations of bowel transit time (e.g., during diarrhea), or stool consistency. On the other hand, plasma/serum-based assays may allow the precise and standardized quantitative expression of viral loads, thus enabling the assessment of disease severity and prognosis. Detection of viral nucleic acids in plasma/serum has been well established for viral load studies for numerous other viruses *(16,17)*. At the beginning of the SARS outbreak, a single report showed the relatively low sensitivity of detecting SARS-CoV RNA in plasma using an ultracentrifugation-based approach, with low concentrations of SARS-CoV detected in the plasma of a patient 9 d after disease onset *(2)*. Subsequently, together with the improvement of viral RNA extraction in which plasma or serum requires no ultracentrifugation, two real-time quantitative RT-PCR assays, one aimed toward the polymerase region and the other toward the nucleocapsid region of the virus genome (**Fig. 1**), were developed for measuring the concentration of SARS-CoV RNA in serum/plasma samples from SARS patients *(13,14)*. In these assays, the absolute calibration curves are constructed by serial dilutions of high-performance liquid chromatography (HPLC)-purified, single-stranded synthetic DNA oligonucleotides specifying the studied amplicons (**Fig. 2**). Previous studies have shown that such single-stranded oligonucleotides reliably mimic the products of the reverse-transcription step and produce calibration curves that are identical to those obtained using T7-transcribed RNA *(18,19)*. The use of such calibration methodology significantly simplifies the process of obtaining a calibration curve when compared with the labor-intensive preparation of calibration curve involving amplicon subcloning and in vitro transcription.

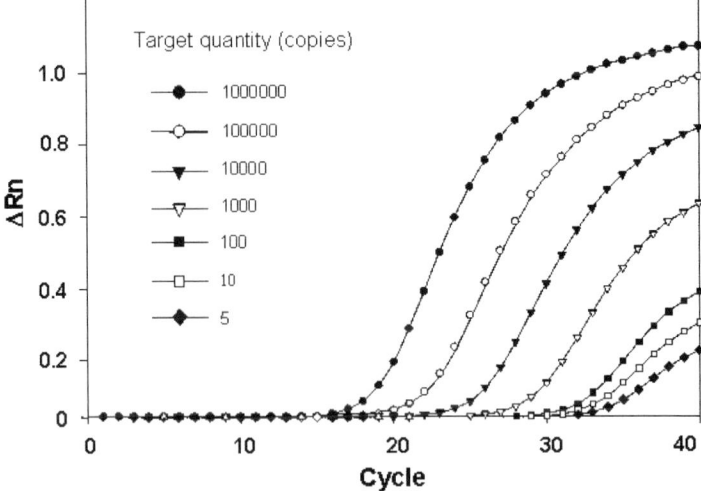

Fig. 2. Detection of severe acute respiratory syndrome (SARS)-coronavirus (CoV) RNA by real-time quantitative reverse-transcription-polymerase chain reaction (PCR) for the nucleocapsid region of the viral genome. An amplification plot of ΔRn, which is the fluorescence intensity over the background (y-axis) against the PCR cycle number (x-axis). Each plot corresponds to a particular input synthetic DNA oligonucleotide target quantity marked by a corresponding symbol.

The sensitivities of the amplification steps of these assays are sufficient to detect 5 copies of the targets in the reaction mixtures, corresponding to 74 copies/mL of serum *(13)*. Using these RT-PCR assays, SARS-CoV RNA was detected in 75–78% of serum samples from SARS patients during the first week of illness *(13)*. In the same study, data showed that median concentrations of serum SARS-CoV RNA in patients who required intensive care unit (ICU) admission during the course of hospitalization were significantly higher than in those who did not require intensive care *(13)* (**Fig. 3**). This quantitative test thus provides the viral load information allowing clinicians to make a prognostic evaluation of the infected individual.

Recent reports revealed that the clinical course of pediatric SARS patients was less severe in comparison with adult SARS patients *(20,21)*. On the whole, the outcomes of pediatric SARS patients were favorable. With the use of the real-time quantitative RT-PCR assay, SARS-CoV RNA has recently been shown to be detectable in the plasma samples of pediatric patients during different stages of SARS (**Fig. 4**) *(14)*. No significant difference in plasma SARS-CoV viral load has been observed between pediatric and adult SARS patients taken within the first week of admission and at d 7 after fever onset *(14)*. Over-

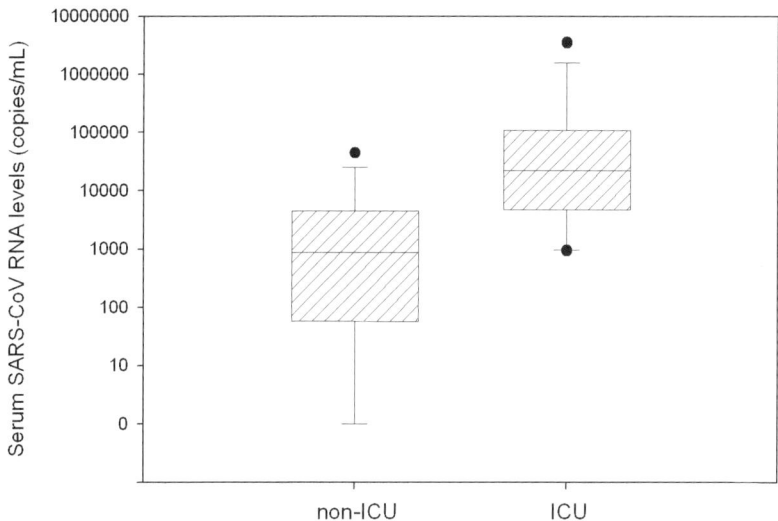

Fig. 3. Serum severe acute respiratory syndrome (SARS)-coronavirus (CoV) RNA levels in SARS patients on the day of hospital admission. Box plot of SARS-CoV RNA levels (common logarithmic scale) in sera of SARS patients requiring and not requiring intensive care unit admission. A real-time quantitative reverse-transcription polymerase chain reaction assay towards the nucleocapsid region of the SARS-CoV genome was used for quantification. The horizontal lines denote the medians. The lines inside the boxes denote the medians. The boxes mark the interval between the 25th and 75th percentiles. The whiskers denote the interval between the 10th and 90th percentiles. The filled circles mark the data points outside the 10th and 90th percentiles.

all, viremia appears to be a consistent feature in both pediatric and adult SARS patients.

The relatively high detection rate of SARS-CoV in plasma and serum during the first week of illness suggests that plasma/serum-based RT-PCR should be incorporated into the routine diagnostic workup of suspected or confirmed SARS patients both in adult and pediatric populations. This approach opens up numerous interesting research opportunities. For example, this assay can be used to monitor the effect or lack of effects of anti-viral agents. Also, it would be valuable to explore the potentially damaging effect of giving steroids at a time when the viral load is still relatively high. We are aware that many of these questions might not be answerable with retrospectively collected samples. Nonetheless, the development of animal models *(5,6)* might allow the testing of some of these hypotheses in a controlled manner.

Patient 1

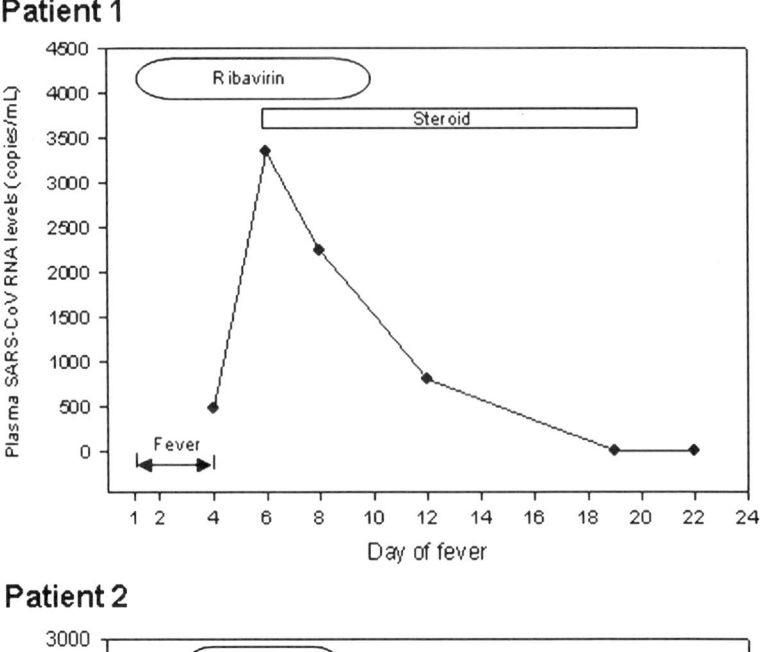

Patient 2

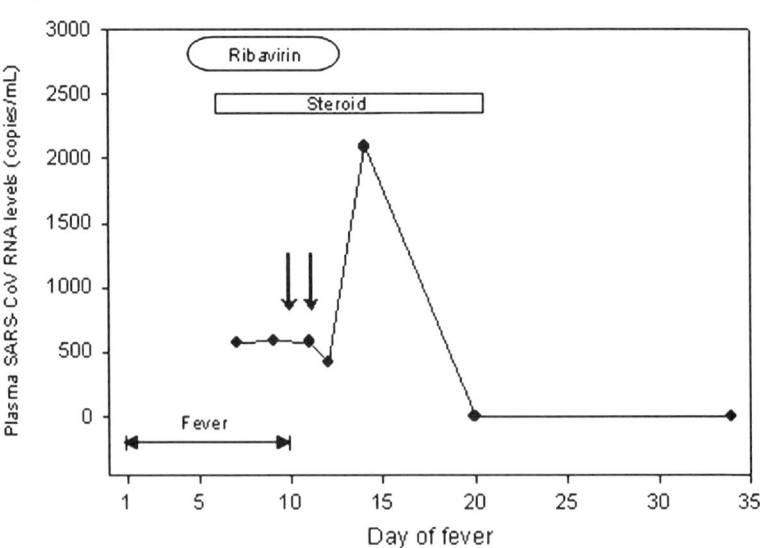

Fig. 4. Serial analysis of plasma severe acute respiratory syndrome (SARS)-coronavirus (CoV) RNA levels in pediatric SARS patients. Plots of plasma SARS-CoV RNA levels (y-axis) against time after the onset of fever (d 1 refers the day of fever onset) (x-axis). The duration of fever and the periods of steroid and ribavirin treatment are indicated for each case. The arrows in patient 2 indicate the time of intravenous methylprednisolone treatment.

In this chapter, the two systems discussed as follows are based on the development of the quantifications of SARS-CoV RNA in serum. The SARSN RT-PCR system was designed to amplify the nucleocapsid region and SARSpol1 system was designed to amplify the polymerase region of the virus. These SARS-CoV RT-PCR systems are useful for the early diagnosis of SARS and can provide viral load information, helping clinicians to make a prognostic evaluation of the infected individual.

2. Materials

2.1. Sample Collection

1. Plain collection tubes for serum collection.
2. RNase Away (Invitrogen, Carlsbad, CA).

2.2. RNA Extraction

1. QIAamp viral RNA Mini Kit (Qiagen, Hilden, Germany).
2. Absolute ethanol.

2.3. Real-Time Quantitative RT-PCR

2.3.1. Amplification Reagents

1. Primers (*see* **Note 1**):
   a. SARSpol1:
      i. Forward: 5'-GAGTGTGCGCAAGTATTAAGTGA-3'.
      ii. Reverse: 5'-TGATGTTCCACCTGGTTTAACA-3'.
   b. SARSN:
      i. Forward: 5'-TGCCCTCGCGCTATTG-3'.
      ii. Reverse: 5'-GGCCTTTACCAGAAACTTTGC-3'.
2. Dual-labeled fluorescent probes (*see* **Note 1**):
   a. SARSpol1:
      5'-(FAM)ATGGTCATGTGTGGCGGCTCACTA(TAMRA)-3'.
   b. SARSN:
      5'-(FAM)TGCTAGACAGATTGAACCAGCTTG(TAMRA)-3',
      where FAM is 6-carboxy-fluorescein; TAMRA is 6-carboxy-tetramethyl-rhodamine.

The amplicon locations of the SARSN and SARSpol1 RT-PCR systems are shown in **Fig. 1**.

3. Calibrators:
   a. Synthetic DNA oligonucleotide corresponding to part of the polymerase gene of SARS-CoV genome, HPLC-purified (Proligos, Singapore) (*see* **Note 2**):
      5'-AACGAGTGTGCGCAAGTATTAAGTGAGATGGTCATGTGTGG
      CGGCTCACTATATGTTAAACCAGGTGGAACATCATCCGG-3'.

    b. Synthetic DNA oligonucleotide corresponding to part of the nucleocapsid gene of SARS-CoV genome, HPLC-purified (Proligos, Singapore) (*see* **Note 2**): 5'-GAAACTGCCCTCGCGCTATTGCTGCTAGACAGATTGAACC AGCTTGAGAGCAAAGTTTCTGGTAAAGGCCAACAA-3'.

4. RNase-free water.
5. EZ r*Tth* RNA PCR reagent set (Applied Biosystems, Foster City, CA).

2.3.2. Instrumentation for Quantitative Analysis

ABI Prism 7700 Sequence Detector (Applied Biosystems).

3. Methods

3.1. Prevention of Contamination

Because of the high sensitivity of RT-PCR-based approaches, strict precautions should be applied to prevent the RT-PCR assay from contamination *(22)*. These precautions include:

1. Aerosol-resistant pipet tips should be used for all liquid handling.
2. Separate areas should be used for the RNA extraction step, the setting up of amplification reactions, the addition of template, and the carrying out of amplification reactions.
3. Real-time PCR approaches obviate the need for post-PCR processing and further reduce the risk of contamination.
4. The assay should include a further level of anticontamination measure in the form of pre-amplification treatment using uracil *N*-glycosylase, which destroys any carried-over uracil containing PCR products *(23)*.
5. Multiple negative water blanks should be included in every analysis so as to detect the possibility of reagent contamination.

3.2. Sample Collection

1. Collect blood samples in plain tubes. To ensure a sufficient amount of serum for RNA analysis, at least 3 mL of blood should be taken for each sample (*see* **Note 3**). The blood samples should be processed as soon as possible to maximize the chance of obtaining good-quality viral RNA. If the processing procedures cannot take place immediately, the blood samples should be stored with extra care (*see* **Note 4**).
2. Centrifuge the blood samples at 1600*g* for 10 min at 4°C.
3. Transfer serum into new tubes.
4. Store the serum at –80°C until RNA extraction (*see* **Note 5**).

3.3. RNA Extraction

The RNA extraction should be performed in a clean and separate area to minimize the chance of cross-contamination.

1. Add 1.12 mL of AVL buffer to 0.28 mL of serum, mix, and incubate at room temperature for 10 min.
2. Add 1.12 mL of absolute ethanol to the mixture and mix.
3. Apply the mixture to an RNeasy mini column and wash the column according to the manufacturer's recommendations.
4. To elute RNA, add 50 μL of AVE buffer onto the silica-gel membrane and incubate for 1 min at room temperature. Centrifuge the RNeasy column for 1 min at 6000*g*.
5. Store the extracted RNA at –80°C until use.

3.4. Real-Time Quantitative RT-PCR

Serum SARS-CoV RNA is quantified by using one-step real-time quantitative RT-PCR *(24)*. In this method, the r*Tth* (*Thermus thermophilus*) DNA polymerase functions both as a reverse transcriptase and a DNA polymerase *(25)* (*see* **Note 7**). Instead of using a two-step RT-PCR approach, one-step RT-PCR, incorporating both the reverse transcription and PCR steps in a single tube, should be used to reduce both hands-on time and the risk of contamination. In this protocol, the quantification of SARS-CoV RNA is described as follows.

3.4.1. Quantification of pol and N Genes of SARS-CoV Genome

1. Prepare calibration curves by serially diluting the synthetic DNA oligonucleotides specifying the amplicons with concentrations ranging from 1.0×10^7 copies to 5 copies (*see* **Note 3**).
2. Set up the RT-PCR reaction mixture for the *Pol* and *N* genes according to **Table 1**.
3. Add 11.25 μL of sample RNA, synthetic DNA oligonucleotides (for calibration curve), or RNase-free water (for negative blanks) into the reaction mixture.
4. Perform the real-time RT-PCR reactions in the ABI Prism 7700 Sequence Detector with cycling conditions shown in **Table 2**.

3.4.2. Data Analysis

Amplification data are analyzed and stored by the Sequence Detection System Software (v1.9; Applied Biosystems). The SARS-CoV RNA concentration is expressed as copies per milliliter of serum (copies/mL). The calculation is shown as follows:

$$C = Q \times \frac{V_{RNA}}{V_{Plasma}}$$

in which C represents the SARS-CoV RNA concentration in serum (copies/mL), Q represents the target quantity (copies/μL) determined by a sequence detector in a PCR, V_{RNA} represents the total volume of RNA obtained after

Table 1
Composition of Reverse-Transcription Polymerase Chain Reaction Mix for Amplification of *Pol* and *N* Genes of Severe Acute Respiratory Syndrome-Coronavirus

Component	Volume for one reaction (µL)	Final concentration
5X TaqMan EZ buffer	5	1X
$Mn(OAc)_2$ (25 mM)	3	3 mM
dATP (10 mM)	0.75	300 µM
dCTP (10 mM)	0.75	300 µM
dGTP (10 mM)	0.75	300 µM
dUTP (20 mM)	0.75	600 µM
Forward primer (10 µM)	0.5	300 nM
Forward primer (10 µM)	0.5	300 nM
Probe (5 µM)	0.5	100 nM
r*Tth* DNA Polymerase (2.5 U/µL)	1	0.1 U/µL
AmpErase uracil-*N*-glycosylase (1 U/µL)	0.25	0.01 U/µL
Total volume	13.75	

Table 2
Cycling Profile for Amplification of *Pol* and *N* Genes of Severe Acute Respiratory Syndrome-Coronavirus

Step	Temperature	Time
Uracil *N*-glycosylase (UNG) treatment	50°C	2 min
Reverse transcription	60°C	30 min
Deactivation of UNG	95°C	5 min
Denaturation	94°C	20 s
40 Cycles		
Annealing/extension	56°C	1 min

extraction (typically 50 µL), V_{Serum} represents the volume of serum extracted (typically 0.28 mL).

The validations of the two real-time RT-PCR systems are as described previously *(13)*.

4. Notes

1. Primers and probes are designed with the use of the *Primer Express®* Software v2.0 (Applied Biosystems). Certain precautions for the design are listed as follows:

    a. The amplicon length should be less than 100 bp, ideally no longer than 80 bp. Short amplicon length is preferable for several reasons, including: (1) the synthetic oligonucleotides specifying the amplicon used as a calibration curve are commercially available with size of up to 100 nucleotides, and (2) amplification with shorter amplicon length is more efficient than that of longer amplicon lengths *(18)*.

    b. To avoid false-positive results arising from co-amplification of genes with high homology, it is necessary to perform a BLASTN search with the primer and probe sequences against the National Center for Biotechnology Information (NCBI) GenBank database. The result of such a search will provide information regarding the specificity of the amplification.

2. Generally, in vitro-transcribed RNA is used as a calibration curve for an absolute RNA quantification. This in vitro-transcribed RNA is usually generated by subcloning the amplicon behind a T7 RNA polymerase promoter in a plasmid vector. However, this procedure is labor-intensive and time-consuming, which is unsuitable when a large number of RT-PCR systems must be constructed in a short period of time. An alternative method can be used to construct a calibration curve, using synthetic single-stranded oligonucleotides corresponding to the amplified sequence. Previous data have shown that such single-stranded oligonucleotides reliably mimic the products of the reverse transcription step and produce calibration curves that are essentially identical to those obtained using T7-transcribed RNA *(18)*.

3. Our studies have revealed that 0.28 mL of plasma is the minimal sample volume for the robust detection of SARS-CoV RNA in serum. Thus, 3 mL of blood samples should be sufficient for testing.

4. When plain blood samples are left unprocessed, their corresponding serum RNA concentrations will fluctuate over time *(26)*. This artifactual fluctuation may be owing to several factors, such as release of RNA from necrotic and/or apoptotic blood cells and the stability of the original and the newly released RNA. To guarantee a reliable serum RNA concentration, we recommend all blood samples be stored at 4°C and be processed as soon as possible.

5. According to our pre-analytical studies, no significant differences in viral RNA concentrations were observed between serum stored at –20°C and –80°C (unpublished data). Nonetheless, we routinely store serum samples at –80°C.

6. The use of one-step, one-enzyme RT-PCR with r*Tth* polymerase has several advantages over the two-enzyme RT-PCR: (1) it has been reported that the *Tth* polymerase is more resistant to inhibitors present in biological specimens than *Taq* polymerase *(27)*; (2) the r*Tth* polymerase is thermostable and thus allows the reverse transcription to perform at high temperature (60°C), thereby minimizing the secondary structures present in the RNA; and (3) as both reverse-transcription and PCR are carried out in a single tube, this reduces both hands-on time and the risk of contamination.

Acknowledgments

This work is supported by a Special Grant for SARS Research (CUHK 4508/03M) from the Research Grants Council of the Hong Kong Special Administrative Region (China).

References

1. Peiris, J. S., Guan, Y., and Yuen, K. Y. (2004) Severe acute respiratory syndrome. *Nat. Med.* **10,** S88–S97.
2. Drosten, C., Gunther, S., Preiser, W., et al. (2003) Identification of a novel coronavirus in patients with severe acute respiratory syndrome. *N. Engl. J. Med.* **348,** 1967–1976.
3. Peiris, J. S., Lai, S. T., Poon, L. L. M., et al. (2003) Coronavirus as a possible cause of severe acute respiratory syndrome. *Lancet* **361,** 1319–1325.
4. Poutanen, S. M., Low, D. E., Henry, B., et al. (2003) Identification of severe acute respiratory syndrome in Canada. *N. Engl. J. Med.* **348,** 1995–2005.
5. Fouchier, R. A., Kuiken, T., Schutten, M., et al. (2003) Aetiology: Koch's postulates fulfilled for SARS virus. *Nature* **423,** 240.
6. Kuiken, T., Fouchier, R. A. M., Schutten M., et al. (2003) Newly discovered coronavirus as the primary cause of severe acute respiratory syndrome. *Lancet* **362,** 263–270.
7. Peiris, J. S., Chu, C. M., Cheng, V. C., et al. (2003) Clinical progression and viral load in a community outbreak of coronavirus-associated SARS pneumonia: a prospective study. *Lancet* **361,** 1767–1772.
8. Tsui, S. K. W., Chim, S. S. C., Lo, Y. M. D., and The Chinese University of Hong Kong Molecular SARS Research Group. (2003) Coronavirus genomic-sequence variations and the epidemiology of the severe acute respiratory syndrome. *N. Engl. J. Med.* **349,** 187–188.
9. Rota, P. A., Oberste, M. S., Monroe, S. S., et al. (2003) Characterization of a novel coronavirus associated with severe acute respiratory syndrome. *Science* **300,** 1394–1399.
10. Marra, M. A., Jones, S. J., Astell, C. R., et al. (2003) The genome sequence of the SARS-associated coronavirus. *Science* **300,** 1399–1404.
11. Poon, L. L. M., Wong, O. K., Luk, W., Yuen, K. Y., Peiris, J. S., and Guan, Y. (2003) Rapid diagnosis of a coronavirus associated with severe acute respiratory syndrome (SARS). *Clin. Chem.* **49,** 953–955.
12. Poon, L. L. M., Chan, K. H., Wong, O. K., et al. (2003) Early diagnosis of SARS coronavirus infection by real time RT-PCR. *J. Clin. Virol.* **28,** 233–238.
13. Ng, E. K. O., Hui, D. S. C., Chan, K. C. A., et al. (2003) Quantitative analysis and prognostic implication of SARS coronavirus RNA in the plasma and serum of patients with severe acute respiratory syndrome. *Clin. Chem.* **49,** 1976–1980.
14. Ng, E. K. O., Ng., P. C., Hon, K. L., et al. (2003) Serial analysis of the plasma concentration of SARS coronavirus RNA in pediatric patients with severe acute respiratory syndrome. *Clin. Chem.* **49,** 2085–2088.

15. Hung, E. C. W., Chim, S. S. C., Chan, P. K., et al. (2003) Detection of SARS coronavirus RNA in the cerebrospinal fluid of a patient with severe acute respiratory syndrome. *Clin. Chem.* **49,** 2108–2109.
16. Zhao, J. R., Bai, Y. J., Zhang, Q. H., Wan, Y., Li, D., and Yan, X. J. (2005) Detection of hepatitis B virus DNA by real-time PCR using TaqMan-MGB probe technology. *World J. Gastroenterol.* **11,** 508–510.
17. Castelain, S., Descamps, V., Thibault, V., et al. (2004) TaqMan amplification system with an internal positive control for HCV RNA quantitation. *J. Clin. Virol.* **31,** 227–234.
18. Bustin, S. A. (2000) Absolute quantification of mRNA using real-time reverse transcription polymerase chain reaction assays. *J. Mol. Endocrinol.* **25,** 169–193.
19. Ng, E. K. O., Tsui, N. B. Y., Lau, T. K., et al. (2003) mRNA of placental origin is readily detectable in maternal plasma. *Proc. Natl. Acad. Sci. USA* **100,** 4748–4753.
20. Chiu, W. K., Cheung, P. C., Ng, K. I., et al. (2003) Severe acute respiratory syndrome in children: Experience in a regional hospital in Hong Kong. *Pediatr. Crit. Care Med.* **4,** 279–283.
21. Hon, K. L., Leung, C. W., Cheng, W. T., et al. (2003) Clinical presentations and outcome of severe acute respiratory syndrome in children. *Lancet* **361,** 1701–1703.
22. Borst, A., Box, A. T. A., and Fluit, A. C. (2004) False-positive results and contamination in nucleic acid amplification assays: suggestions for a prevent and destroy strategy. *Eur. J. Clin. Microbiol. Infect. Dis.* **23,** 289–299.
23. Longo, M. C., Berninger, M. S., and Hartley, J. L. (1990) Use of uracil DNA glycosylase to control carry-over contamination in polymerase chain reactions. *Gene* **93,** 125–128.
24. Gibson, U. E., Heid, C. A., and Williams, P. M. (1996) A novel method for real time quantitative RT-PCR. *Genome Res.* **6,** 995–1001.
25. Myers, T. W. and Gelfand, D. H. (1991) Reverse transcription and DNA amplification by a Thermus thermophilus DNA polymerase. *Biochemistry* **30,** 7661–7666.
26. Tsui, N. B. Y., Ng, E. K. O., and Lo, Y. M. D. (2002) Stability of endogenous and added RNA in blood specimens, serum, and plasma. *Clin. Chem.* **48,** 1647–1653.
27. Poddar, S. K., Sawyer, M. H., and Connor, J. D. (1998) Effect of inhibitors in clinical specimens on Taq and Tth DNA polymerase-based PCR amplification of influenza A virus. *J. Med. Microbiol.* **47,** 1131–1135.

16

Genomic Sequencing of the Severe Acute Respiratory Syndrome-Coronavirus

Stephen S. C. Chim, Rossa W. K. Chiu, and Y. M. Dennis Lo

Summary

The polymerase chain reaction (PCR), which can exponentially replicate a target DNA sequence, has formed the basis for the sensitive and direct examination of clinical samples for evidence of infection. During the epidemic of severe acute respiratory syndrome (SARS) in 2003, PCR not only offered a rapid way to diagnose SARS-coronavirus (SARS-CoV) infection, but also made the molecular analysis of its genomic sequence possible. Sequence variations were observed in the SAR-CoV obtained from different patients in this epidemic. These unique viral genetic signatures can be applied as a powerful molecular tool in tracing the route of transmission and in studying the genome evolution of SARS-CoV. To extract this wealth of information from the limited primary clinical specimens of SARS patients, we were presented with the challenge of efficiently amplifying fragments of the SARS-CoV genome for analysis. In this chapter, we will discuss how we managed to accomplish this task with our optimized protocols on reverse-transcription, nested PCR amplification, and DNA cycle sequencing. We will also discuss the sequence variations that typified some strains of SARS-CoV in the different phases during this epidemic. PCR amplification of the viral sequence and genomic sequencing of these critical sequence variations of re-emerging SARS-CoV strains would give us quick insights into the virus.

Key Words: SARS coronavirus; viral RNA extraction; reverse-transcription PCR; sequencing; genomic sequence variation.

1. Introduction

Severe acute respiratory syndrome-coronavirus (SARS-CoV), the etiologic agent of SARS *(1–3)*, is a virus that was unknown to us before the SARS epidemic. The concerted efforts of researchers have promptly elucidated its genetic code. The genome of SARS-CoV is a 29,727-nucleotide, polyadenylated RNA. The genomic organization is typical of coronaviruses, having the characteristic gene order (5'-*polymerase [Orf1ab], spike [S], envelope [E], mem-*

From: *Methods in Molecular Biology, vol. 336: Clinical Applications of PCR*
Edited by: Y. M. D. Lo, R. W. K. Chiu, and K. C. A. Chan © Humana Press Inc., Totowa, NJ

brane [*M*], and *nucleocapsid* [*N*]-3') and short untranslated regions at both termini *(4,5)*.

With this sequence information, rapid PCR-based molecular diagnostic tests of SARS-CoV infection were designed *(1,6–10)*. Besides offering molecular diagnosis and quantitative measurement of viral load, PCR-based technologies have also been exploited to amplify the genomic fragments of SARS-CoV for sequence analysis. The high sensitivity and specificity of PCR has made this genomic sequence analysis possible even for uncultured clinical specimens. Unlike the conventional microbiological methods, PCR-based technologies may not require viral culture, which could introduce culture-derived artifacts in the genomic sequence. The specific PCR primers selectively amplify SARS-CoV sequences from the background of other nucleic acid sequences contributed by the patient or other microbes. Moreover, the PCR-based method is versatile in terms of the type of clinical specimens. In our hands, we have successfully analyzed the SARS-CoV genome directly from uncultured samples of serum, nasopharyngeal aspirate, and stools *(11)*. This obviates any concern about the poor or even unsuccessful viral culture of the precious clinical specimens. The risk in handling large-volume and hazardous viral culture could also be avoided.

Genomic sequence variations were observed in the SARS-CoV obtained from different patients in this epidemic. Based on these sequence variations, most of the isolates are typified by two groups: isolates obtained from patients who were epidemiologically linked to the Metropole Hotel in Hong Kong, and those who were not *(3,12,13)*. For example, there are seven sequence variations that can distinguish isolate CUHK-Su10, which is linked to the Metropole Hotel, from isolate CUHK-W1, which is not linked to this hotel case cluster (**Table 1**). Among them, four variations at nucleotide positions 17564, 21721, 22222, and 27827 (according to the Tor2 sequence in GenBank, accession no. AY274119 *[5]*) were suggested by The Chinese SARS molecular epidemiology consortium *(14)* as part of a haplotype configuration that marks the different phases of a tri-phasic SARS epidemic in Guangdong Province of China. CUHK-W1 carried a haplotype G:A:C:C that typified the middle phase. Notably, the same haplotype was observed in CUHK-L2, which was one of the earliest confirmed case of SARS in Hong Kong, having been documented even before any report of the hotel case cluster *(15)*. CUHK-Su10 carried a haplotype T:T:T:T that typified the late phase, marked by the hotel case cluster that spread the virus to many other parts of the world.

Genomic sequence variations in SARS-CoV have also revealed the route of infection from within communities and across cities. For instance, compared with isolate CUHK-Su10, two mutations, T3852C and C11493T, first appeared

Table 1
Comparison of the Sequences of Two Strains of Severe Acute Respiratory Syndrome (SARS)-Coronavirus Isolated From Patients in Hong Kong at the Beginning of the Epidemic[a]

Nucleotide position	CUHK-Su10	CUHK-W1
9404	T	C
9479	T	C
17564[b]	T	G
19064	A	G
21721[b]	G	A
22222[b]	T	C
27827[b]	T	C

[a]Sequence variations at seven positions between the two viral strains (CUHK-Su10 and CUHK-W1) are indicated. The nucleotide positions are numbered according to the sequence of GenBank accession number AY274119.

[b]Part of the haplotype suggested by The Chinese SARS molecular epidemiology consortium for distinguishing the early, middle, and late phase of the SARS epidemic in 2003.

in isolates CUHK-AG01, CUHK-AG02, CUHK-AG03 (GenBank accession numbers AY345986, AY345987, AY345988) obtained from patients involved in the Amoy Gardens outbreak in Hong Kong *(11)*. Later, these two genetic fingerprints appeared in 10 completely sequenced Taiwanese isolates *(16)*.

Interestingly, toward the end of the epidemic, another type of fingerprint was found by PCR-based method. A variant of the SARS-CoV with a 386-nucleotide deletion was reported in a cluster of patients that seem to be epidemiologically related *(17)*. Most of the cases were part of a documented outbreak in the North District Hospital in Hong Kong.

We have illustrated that sequence variations among different isolates have a remarkable epidemiological correlation. Thus, PCR amplification followed by sequencing is a powerful tool in tracing the route of transmission. The sequence information may provide objective support to epidemiological investigations. Moreover, in the event that SARS re-emerges, one could quickly gain important insight into the origin and evolution status of the SARS-CoV simply by sequencing the critical sequence variations of the genome, as exemplified here. However, to extract this wealth of information from the limited primary clinical SARS specimens, we need very sensitive and efficient protocols to efficiently amplify fragments of the SARS-CoV genome for analysis.

2. Materials

2.1. RNA Extraction

1. QIAamp viral RNA Mini Kit (Qiagen, Hilden, Germany).
2. Absolute ethanol.

2.2. Reverse-Transcription

1. Superscript III RNase H⁻ reverse transcriptase (Invitrogen, Carlsbad, CA).
2. Random hexamers (Applied Biosystems, Foster City, CA).
3. 5X First-strand synthesis buffer: 250 mM Tris-HCl, pH 8.3, 375 mM KCl, 15 mM MgCl$_2$ (Invitrogen).
4. RNasin RNase inhibitor (Promega, Madison, WI).
5. dNTP (Invitrogen).
6. 0.1 M Dithiothreitol (DTT).
7. RNase-free water (Promega).

2.3. PCR Amplification

1. Advantage cDNA Polymerase mix and buffer (BD Biosciences Clontech, Palo Alto, CA).
2. dNTP.
3. PCR primers.
4. Distilled water.

2.4. Genomic Sequencing

1. BigDye Terminator v1.1 Cycle Sequencing Kit (Applied Biosystems).
2. ABI Prism 3100 Genetic Analyzer (Applied Biosystems).
3. DYEnamic ET Dye Terminator Kit (GE Healtcare-Biosciences, Little Chalfont, UK).
4. MegaBACE 1000 Sequencing System (GE Healthcare-Biosciences).

2.5. Sequence Analysis and Comparison

SeqScape software (Applied Biosystems).

3. Methods

3.1. Precautions Against Potential Contamination

Genomic sequencing involves PCR amplification, which produces numerous copies of the target DNA, and cycle sequencing, which requires the pipetting and manipulation of PCR products. These steps could easily contaminate the laboratory environment with amplified products. Such contamination problems would affect the interpretation of sequencing results, and adversely affect the performance of diagnostic tests designed to detect the same viral sequences. Hence, extreme care should be taken to avoid contamination. We suggest the following precautions:

1. Perform RNA extraction, PCR amplification, and genome sequencing in different laboratories, or at least in separate and dedicated compartments of the same laboratory.
2. Transfer reagents and samples only with aerosol-resistant pipet tips.
3. Prepare the PCR reagent master mix in a hood dedicated for this purpose. A set of clean gloves and dedicated lab gown should be worn in this area. Illuminate the hood with ultraviolet before and after use.
4. Any steps that involve the handling of cDNA, primary and secondary PCR products (including addition of DNA templates in assembling the PCR), electrophoresis, and cycle sequencing should be performed in a dedicated area far away from any PCR reagents. A separate lab gown and set of gloves should be worn in this area.
5. Discard all pipet tips that contacted DNA with extreme care. Use a double bag for disposal.
6. Include multiple negative PCR controls in each amplification to monitor for environmental contamination.

3.2. RNA Extraction

1. Prepare AVL lysis buffer and AW1 and AW2 wash buffers according to manufacturer's (Qiagen) instructions (*see* **Note 1**).
2. In a biosafety level 2 (or above) containment laboratory, lyse 0.28 mL (1 vol) of viral culture by adding 1.12 mL (4 vol) of AVL buffer, mixing and incubating at room temperature for 10 min. Direct clinical samples, e.g., serum, nasopharyngeal aspirates, and stools, can also be used (*see* **Note 2**).
3. Add 1.12 mL of absolute ethanol to the mixture. Pulse-vortex for 15 s.
4. Load the mixture to QIAamp spin column and wash the column according to the manufacturer's instructions.
5. Add 60 μL of RNase-free water onto the membrane and incubate for 1 min at room temperature. Centrifuge the spin column for 1 min at 6000*g*.
6. Quantify a small aliquot of the extracted viral RNA yield by real-time quantitative reverse-transcription (RT)-PCR *(9)* (*see* **Note 3**).
7. Store the extracted RNA at –80°C.

3.3. Reverse-Transcription

1. Prewarm two thermocycler blocks with heated lid at 72 and 25°C, respectively.
2. Mix 1 μL (50 pmol) random hexamer with 10 μL RNA in a 0.5-mL tube. Denature at 72°C for 10 min (*see* **Note 4**).
3. During this period, assemble the reaction mix in another tube on ice according to **Table 2** using SuperScript III RNase H⁻ Reverse Transcriptase (*see* **Note 5**).
4. After denaturation, snap-cool the RNA-primer mixture on ice for 1 min. Briefly spin the tubes. Add the reaction mix prepared in **step 2** to the RNA-primer mixture to make up a total reaction volume of 20 μL. Mix by pipetting gently up and down.

Table 2
Composition of Reaction Mix for Reverse-Transcription of Severe Acute Respiratory Syndrome-Coronavirus RNA

Component	Volume for one reaction (µL)	Final concentration
5X first strand buffer	4	1X
dNTP mix (10 mM each dATP, dGTP, dCTP, and dTTP at neutral pH)	1	0.5 mM each
0.1 M dithiothreitol 1	5 mM	
RNasin RNase inhibitor (40 U/µL)	1	2 U/µL
SuperScript III reverse transcriptase (200 U/µL)	2	20 U/µL
Total volume	9	

Table 3
Composition of Reaction Mix for Polymerase Chain Reaction Amplification

Component	Volume for one reaction (µL)	Final concentration
Distilled water	19.5	
Advantage PCR buffer (10X)	2.5	1X
dNTP mix (10 mM each dATP, dGTP, dCTP, and dTTP at neutral pH)	0.5	200 µM each
cDNA polymerase mix (50X)	0.5	1X
Total volume	23.0	

5. Immediately transfer the tube from ice to the prewarmed 25°C thermocycler block for a 5-min incubation. Prewarm the other thermocycler block at 55°C.
6. Transfer the tube to the prewarmed 55°C thermocycler block for a 1-h incubation.
7. Heat inactivate at 72°C for 15 min.
8. Add 1 µL (2 U) of RNase H and incubate at 37°C for 20 min to remove RNA complementary to the cDNA.
9. Dilute the product two- to fivefold with distilled water. Store at –20°C before use.

3.4. Primary PCR Amplification

1. Inside a hood dedicated for setting up PCR, assemble the PCR master mix for the 50 reactions according to **Table 3** with cDNA polymerase mix (*see* **Note 6**) in a

final reaction volume of 25 μL. Add 50 aliquots of 23 μL into a 96-well PCR microplate.

2. Add 5 pmol each of forward (PCR-F) and reverse (PCR-R) series of primers for each of the 50 reactions amplifying the overlapping amplicons that cover the whole SARS-CoV genome (*see* **Note 7**). The primer sequences are shown in **Table 4**.

3. In an area separate from the hood dedicated for PCR, add 1 μL of diluted reverse-transcribed products.

4. Commence with PCR in a thermocycler with initial denaturation at 95°C for 1 min and 35 cycles of 95°C for 0.5 min, 55°C for 0.5 min, 68°C for 1.5 min, and a final extension at 68°C for 10 min.

3.5. Secondary PCR Amplification

1. Inside a hood dedicated for setting up PCR, assemble the PCR master mix for the 50 reactions according to **Table 3** in a final reaction volume of 25 μL. Add 50 aliquots of 23 μL into a new 96-well PCR microplate.

2. Add 5 pmol each of forward (PCR-F) and reverse (BSEQ-R) series of primers for each of the 50 semi-nested PCR reactions. The primer sequences are shown in **Table 4**.

3. In an area separate from the hood dedicated for PCR, add 1 μL of the corresponding primary PCR product.

4. Commence PCR in a thermocycler with initial denaturation at 95°C for 1 min and 35 cycles of 95°C for 0.5 min, 55°C for 0.5 min, 68°C for 1.5 min, and a final extension at 68°C for 10 min.

5. Electrophorese 5 μL of the secondary PCR product in a 2% agarose gel to verify the success of the PCR amplification. Estimate the amount of PCR product by comparison to DNA marker. Only products with single band should be used for sequencing.

3.6. Cycle Sequencing

Perform sequencing reaction based on the dideoxy dye terminator method, according to manufacturers' instructions:

1. Separate from the hood dedicated for PCR, assemble the cycle sequencing reaction with ASEQ-F, BSEQ-F, ASEQ-R, and BSEQ-R series of oligonucleotides as sequencing primers for each of the amplicon, and with 2–5 ng of secondary PCR product as sequencing template (*see* **Note 8**).

2. Commence with cycle sequencing reaction in a thermocycler.

3. Purify the extension products with either spin column purification or ethanol precipitation. Mix or resuspend the DNA in formamide solution according to the manufacturer's instructions.

4. Denature the purified extension products at 95°C for 5 min, snap-cool on ice, and load onto the automated capillary DNA sequencer for injection.

Table 4
Primer Sequences

01PCR-F	CTACCCAGGAAAAGCCAACCAACCT
01ASEQ-F	AAAGCCAACCAACCTCGATC
01ASEQ-R	AAGTGCCATTTTTGAGGTGT
01BSEQ-F	TTGCCTGTCCTTCAGGTTAG
01BSEQ-R	GTCACCTAAGTCATAAGACT
01PCR-R	TGCCAAGCTCGTCACCTAAGTCATA
02PCR-F	TACCGCAATGTTCTTCTTCGTAAGA
02ASEQ-F	TTCTTCTTCGTAAGAACGGT
02ASEQ-R	GCTCGTAGCTCTTATCAGAG
02BSEQ-F	CAACTTGATTACATCGAGTC
02BSEQ-R	TTCAGTGCCACAATGTTCAC
02PCR-R	TAACTAAATTTTCAGTGCCACAATG
03PCR-F	TCTACCTTGATGGGGTGTAATCATT
03ASEQ-F	TGAAATCTAATCATTGCGAT
03ASEQ-R	GGAGATCCTCATTCAAGGTC
03BSEQ-F	AATAAGCGTGCCTACTGGGT
03BSEQ-R	AATTGATCTGATAACACCAG
03PCR-R	TGCGCGCAAAAATTGATCTGATAAC
04PCR-F	AAAGGTGCTTGGAACATTGGACAAC
04ASEQ-F	GGAACATTGTACAACAGAGA
04ASEQ-R	ATTTGAGAATCTCCCAAGCA
04BSEQ-F	GGCACTACTGTTGAAAAACT
04BSEQ-R	ATGTGAATCACCTTCAAGAA
04PCR-R	GTACTGTGTCATGTGAATCACCTTC
05PCR-F	CGTCAGTGTATACGTGGCAAGGAGC
05ASEQ-F	TACGTGGCAAGGAGCAGCTG
05ASEQ-R	CAACACGTTCATCAAGCTCA
05BSEQ-F	GGTGCACCAATTAAAGGTGT
05BSEQ-R	ACAGGTTTCATCAATTTCTT
05PCR-R	ACTCATGTTCACAGGTTTCATCAAT
06PCR-F	TCATCACGTATGTATTGTTCCTTTT
06ASEQ-F	TGTATTGTTCCTTTTACCCT
06ASEQ-R	CTGCTACACCACCACCATGT
06BSEQ-F	ATTAAATGTGTTGACATCGT
06BSEQ-R	AACCGTCTGCACGCACACTT
06PCR-R	CCTGTGTACGAACCGTCTGCACGCA
07PCR-F	TCACAGGACATCTTACTTGCACCAT
07ASEQ-F	TCTTACTTGCACCATTGTTG
07ASEQ-R	CTTCACCTCTAAGCATGTTC
07BSEQ-F	ACACTGGAAGAAACTAAGTT

(continued)

Table 4 (*Continued*)
Primer Sequences

07BSEQ-R	TACAGTTCCTAGAATCTCTT
07PCR-R	AATTCCAGGATACAGTTCCTAGAAT
08PCR-F	GCTAAGACTGCTCTTAAGAAATGCA
08ASEQ-F	CTCTTAAGAAATGCAAATCT
08ASEQ-R	CTACGGCAGGAGCTTTAAGA
08BSEQ-F	AATGAGCCGCTTGTCACAAT
08BSEQ-R	CACTTTTATAGTCTTAACCT
08PCR-R	CAGTTGTGAACACTTTTATAGTCTT
09PCR-F	CCCGTCGAGTTTCATCTTGACGGTG
09ASEQ-F	TTCATCTTGACGGTGAGGTT
09ASEQ-R	AATTGTTATCAGCCCATTTA
09BSEQ-F	AGTTTTCTTGGTAGGTACAT
09BSEQ-R	ATACATCACAGCTTCTACAC
09PCR-R	GAGTACCCATATACATCACAGCTTC
10PCR-F	TTGGAATCTGCAAAGCGAGTTCTTA
10ASEQ-F	CAAAGCGAGTTCTTAATGTG
10ASEQ-R	TGTAAGATGTTTCCTTGTAG
10BSEQ-F	GCTAAGGAGACCCTCTATCG
10BSEQ-R	AATAGCCACTACATCGCCAT
10PCR-R	GTCTATAGTCAATAGCCACTACATC
11PCR-F	TTAAATCAAATGACAGGCTTCACAA
11ASEQ-F	TGACAGGCTTCACAAAGCCA
11ASEQ-R	TGCCTACAACTTCGGTAGTT
11BSEQ-F	TGTGAAAGTCAACAACCCAC
11BSEQ-R	AGGCATATAATTGTTAAACA
11PCR-R	TAAACACATAAGGCATATAATTGTT
12PCR-F	TATGTCAAACCATTCTTAGGACAAG
12ASEQ-F	CATTCTTAGGACAAGCAGCA
12ASEQ-R	ACGAATTAAGATACAATTCT
12BSEQ-F	GGTTCTCTAATCTGTGTAAC
12BSEQ-R	ACTAATGATAAACCACATGA
12PCR-R	TTTGTACAATACTAATGATAAACCA
13PCR-F	TTAGGTCTTTCAGCTATAATGCAGG
13ASEQ-F	CAGCTATAATGCAGGTGTTC
13ASEQ-R	ACAAATCACGAGCAACTTCA
13BSEQ-F	GGCCGTGGCTTCTGCAAGAC
13BSEQ-R	CAGAATAGGTTGGCACATCA
13PCR-R	GGTCAAGCAACAGAATAGGTTGGCA
14PCR-F	ATAGTTTTTGATGGCAAGTCCAAAT
14ASEQ-F	ATGGCAAGTCCAAATGCGAC

(*continued*)

Table 4 (*Continued*)
Primer Sequences

14BSEQ-F	GTTGTTGATACCGATGTTGA
14BSEQ-R	AAAACAAGTACTAACAATCT
14ASEQ-R	TGACAGTTGTAACAATTTCA
14PCR-R	GCATAAGTTTAAAACAAGTACTAAC
15PCR-F	AGACTAACTTGTGCTACAACTAGAC
15ASEQ-F	GTGCTACAACTAGACAGGTT
15BSEQ-F	GCGTGGTGGTTCATACAAAAA
14BSEQ-R	AAAACAAGTACTAACAATCT
14ASEQ-R	TGACAGTTGTAACAATTTCA
14PCR-R	GCATAAGTTTAAAACAAGTACTAAC
15PCR-F	AGACTAACTTGTGCTACAACTAGAC
15ASEQ-F	GTGCTACAACTAGACAGGTT
15BSEQ-F	CGTGGTGGTTCATACAAAAA
15BSEQ-R	CTCCAGGTAAGTGTTAGGAA
15ASEQ-R	GCACAGTACCCGGTAAGCCA
15PCR-R	TAACAGAACCCTCCAGGTAAGTGTT
16PCR-F	GGTTCTATTTCTTATGGTGAGCTTC
16ASEQ-F	CTTATAGTGAGCTTCGTCCA
16ASEQ-R	ACTCACCAAAAACACGTCTG
16BSEQ-F	TTAGATGTGTCTGCTTCAGT
16BSEQ-R	AACTCCATTAAACATGACTC
16PCR-R	TACTAAATGTAACTCCATTAAACAT
17PCR-F	ACAGCAATCTATGTATTCTGTATTT
17ASEQ-F	ATGTATTCTGTATTTCTCTG
17ASEQ-R	CTGACGGGAATGCCATTTTC
17BSEQ-F	TTTAGCAACTCAGGTGCTGA
17BSEQ-R	TTGGATACGGACAAATTTAT
17PCR-R	TTTGACCAGGTTGGATACGGACAAA
18PCR-F	ATTGGCCATTCTATGCAAAATTGTC
18ASEQ-F	CTATGCAAAATTGTCTGCTT
18ASEQ-R	CAGCATACAGCCATGCCAAA
18BSEQ-F	GAAGGTAAATTCTATGGTCC
18BSEQ-R	ACCTTGGAAGGTAACACCAG
18PCR-R	TCTTGAACTTACCTTGGAAGGTAAC
19PCR-F	GGTCGTACTATCCTTGGTAGCACTA
19ASEQ-F	TCCTTGGTAGCACTATTTTA
19ASEQ-R	AGCTAGTGTCAGCCAATTCA
19BSEQ-F	TTACCTTCTCTTGCAACAGT
19BSEQ-R	AAATAACAATGGGTAATACT
19PCR-R	TGCCAGTAATAAATAACAATGGGTA

(*continued*)

Table 4 (*Continued*)
Primer Sequences

20PCR-F	ACCTCTAACTATTCTGGTGTCGTTA
20ASEQ-F	ATTCTGGTGTCGTTACGACT
20ASEQ-R	AGAGCAGTACCACAGATGTG
20BSEQ-F	AACATTAAGTTGTTGGGTAT
20BSEQ-R	ATTAGCTACAGCCTGCTCAT
20PCR-R	CAGAATCACCATTAGCTACAGCCTG
21PCR-F	CTTCAGGCTATTGCTTCAGAATTTA
21ASEQ-F	TTGCTTCAGAATTTAGTTCT
21ASEQ-R	CAACAACCATGAGTTTGGCT
21BSEQ-F	GATAATGATGCACTTAACAA
21BSEQ-R	GGCAAGTGCATTGTCATCAG
21PCR-R	TGTTATAGTAGGCAAGTGCATTGTC
22PCR-F	AATAATGAACTGAGTCCAGTAGCAC
22ASEQ-F	TGAGTCCAGTAGCACTACGA
22ASEQ-R	CTGCAAAAGCACAGAAGGAA
22BSEQ-F	ATGGTGCTGGGCAGTTTAGC
22BSEQ-R	ACAGACTGTGTTTCTAAGTG
22PCR-R	CGCAGACGGTACAGACTGTGTTTCT
23PCR-F	GGATTCTGTGACTTGAAAGGTAAGT
23ASEQ-F	GACTTGAAAGGTAAGTACGTC
23ASEQ-R	TGTTGGTAGTTAGACATAGT
23BSEQ-F	TAAAAACTAATTGCTGTCGC
23BSEQ-R	CTAAGTTAGCATATACGCGT
23PCR-R	ACACGCTCACCTAAGTTAGCATATA
24PCR-F	ACAATTGCTGTGATGATGATTATTT
24ASEQ-F	TGATGATGATTATTTCAATA
24ASEQ-R	AGACAAAGTCTCTCTTCCGT
24BSEQ-F	GGGCATTGGCTGCTGAGTCC
24BSEQ-R	CTGGATCAGCAGCATACACT
24PCR-R	GCATGCATAGCTGGATCAGCAGCAT
25PCR-F	GTGAGTTAGGAGTCGTACATAATCA
25ASEQ-F	AGTCGTACATAATCAGGATG
25ASEQ-R	CAATCAAAGTATTTATCAAC
25BSEQ-F	CTATCAGTGATTATGACTAT
25BSEQ-R	TTGACTTCAATAATTTCTGA
25PCR-R	GTGGCGGCTATTGACTTCAATAATT
26PCR-F	GTGCAAAGAATAGAGCTCGCACCGT
26ASEQ-F	TAGAGCTCGCACCGTAGCTG
26ASEQ-R	GATGATGTTCCACCTGGTTT
26BSEQ-F	CACACCGTTTCTACAGGTTA

(*continued*)

Table 4 (*Continued*)
Primer Sequences

26BSEQ-R	TGCTAGCTACTAAACCTTGA
26PCR-R	AAGTTCTTAATGCTAGCTACTAAAC
27PCR-F	TGCGTAAACATTTCTCCATGATGAT
27ASEQ-F	TTTCTCCATGATGATTCTTT
27ASEQ-R	TGAAAGACATCAGCATACTC
27BSEQ-F	AAACAGATGGTACACTTATG
27BSEQ-R	GATTAACAGACAACACTAAT
27PCR-R	CAAACATAGGGATTAACAGACAACA
28PCR-F	GTGCCTGTATTAGGAGACCATTCCT
28ASEQ-F	TAGGAGACCATTCCTATCTT
28ASEQ-R	CCATATGACAGCTTAAATGT
28BSEQ-F	CTGGCGATTACATACTTGCC
28BSEQ-R	CATAGTGCTCTTGTGGCACT
28PCR-R	GTAATTCTCACATAGTGCTCTTGTG
29PCR-F	ACAAGTTGAATGTTGGTGATTACTT
29ASEQ-F	TGTTGGTGATTACTTTGTGT
29ASEQ-R	GAATTCACTTTGAATTTATC
29BSEQ-F	TATGTGAAAAGGCATTAAAA
29BSEQ-R	CAGCAGGACAACGGCGACAA
29PCR-R	TCAACAATTTCAGCAGGACAACGGC
30PCR-F	TAGAACCAGAATATTTTAATTCAGT
30ASEQ-F	ATATTTTAATTCAGTGTGCA
30ASEQ-R	GTAGTTTGTGTGAATATGAC
30BSEQ-F	CACAGAACGCTGTAGCTTCA
30BSEQ-R	GTATGCCTGGTATGTCAACA
30PCR-R	ATGTCCTTTGGTATGCCTGGTATGT
31PCR-F	CTGGTCTTCATCCTACACAGGCACC
31ASEQ-F	TCCTACACAGGCACCTACAC
31ASEQ-R	AGATGTTTAAACTGGTCACC
31BSEQ-F	ACTTAGTAGCTGTACCGACT
31BSEQ-R	GCTGAACATCAATCATAAAT
31PCR-R	AAGCCCCACTGCTGAACATCAATCA
32PCR-F	GCTTTTCTACTTCATCAGATACTTA
32ASEQ-F	AAGTAGTCTATGAATACGGA
32ASEQ-R	TTCCATTCTACTTCAGCCTG
32BSEQ-F	TTGTGAAGTCTGCATTGCTT
32BSEQ-R	AAAAGAAAGGCAATTGCTTT
32PCR-R	TCAGAATAGTAAAAGAAAGGCAATT
33PCR-F	GTGGTAGTTTGTATGTGAATGGGCA
33ASEQ-F	GTATGTGAATAAGCATGCAT

                                                    (*continued*)

Table 4 (*Continued*)
Primer Sequences

33ASEQ-R	GCTTCGCCGGCGTGTCCATC
33BSEQ-F	CTTATAACCTGTGGAATACA
33BSEQ-R	GTGAAGAACAAGCACTCTCA
33PCR-R	AAGACAGTAAGTGAAGAACAAGCAC
34PCR-F	AAAGAGAAGCCCCAGCACATGTATC
34ASEQ-F	CCCAGCACATGTATCTACAA
34ASEQ-R	ATGAATTCATCCATAGCGAG
34BSEQ-F	TGCCTGAAACCTACTTTACT
34BSEQ-R	TGACCACTTTTGAAATCACT
34PCR-R	ATTGTAACCTTGACCACTTTTGAAA
35PCR-F	AATGTGTGTGTTCTGTGATTGATCT
35ASEQ-F	TTCTGTGATTGATCTTTTAC
35ASEQ-R	TTATCAGAGCCAGCACCAAA
35BSEQ-F	ATGTCGCAAAGTATACTCAA
35BSEQ-R	CTGTTATCTTTACAGCTATA
35PCR-R	CAAGAATGCTCTGTTATCTTTACAG
36PCR-F	ATGACTCTAAAGAAGGGTTTTTCAC
36ASEQ-F	AGAAGGGTTTTTCACTTATC
36ASEQ-R	TCCAGAAGAGAATAAATCAT
36BSEQ-F	CACTCTTTGACATGAGCAAA
36BSEQ-R	TAGTATGAAACCCTGTAACA
36PCR-R	GTATGATTAATAGTATGAAACCCTG
37PCR-F	ATCCTGATGAAATTTTTAGATCAGA
37ASEQ-F	AATTTTTAGATCAGACACTC
37ASEQ-R	TTTTCTGAAACATCAAGCGA
37BSEQ-F	AACCCATGGGTACACAGACA
37BSEQ-R	TTGTACCATTTTCATCATAC
37PCR-R	GCATCTGTGATTGTACCATTTTCAT
38PCR-F	AAGACATTTGGGGCACGTCAGCTGC
38ASEQ-F	GGGCACGTCAGCTGCAGCCT
38ASEQ-R	TTCAACTTAGTGGCAGAAAC
38BSEQ-F	GAAAAAAAATTTCTAATTGT
38BSEQ-R	GAGCAGGTGGGGTGCAAGGT
38PCR-R	TAACAATTAAGAGCAGGTGGGGTGC
39PCR-F	ATCTTAGACATGGCAAGCTTAGGCC
39ASEQ-F	TGGCAAGCTTAGGCCCTTTG
39ASEQ-R	GGTGAAATGTCTAATATTTC
39BSEQ-F	CAAAGAGATTTCAACCATTT
39BSEQ-R	TTTGGCTAGTACTACGTAAT
39PCR-R	ACAATAGATTTTTGGCTAGTACTAC

(*continued*)

Table 4 (*Continued*)
Primer Sequences

40PCR-F	TCGACACTTCTTATGAGTGCGACAT
40ASEQ-F	TTATGAGTGCGACATTCCTA
40ASEQ-R	GTTGGGGTTTTGTACATTTG
40BSEQ-F	GCACACAACTAAATCGTGCA
40BSEQ-R	CACCAAATGTCCATCCAGCA
40PCR-R	GCAGGCCAGCACCAAATGTCCATC
41PCR-F	TGTTGCCACCTCTGCTCACTGATGA
41ASEQ-F	TCTGCTCACTGATGATATGA
41ASEQ-R	ATTTGTACCTCCGCCTCGAC
41BSEQ-F	ACACACTTGTTAAACAACTT
41BSEQ-R	GGAAGTATGCTTTGCCTTCA
41PCR-R	CCTTCACGAGGGAAGTATGCTTTGC
42PCR-F	TTGTCTTCCTACATGTCACGTATGT
42ASEQ-F	ACATGTCACGTATGTGCCAT
42ASEQ-R	AAATTTTTAGCGACCTCATT
42BSEQ-F	CATCACCAGATGTTGATCTT
42BSEQ-R	ATTGATCCAAGAGTAAAAAA
42PCR-R	CTGTGCAGTAATTGATCCAAGAGTA
43PCR-F	ACTCTGAGCCAGTTCTCAAGGGTGT
43ASEQ-F	AGTTCTCAAGGGTGTCAAAT
43ASEQ-R	TTGTAGAAAATATATCAAGG
43BSEQ-F	ACTGCTGCTATTTGTTACCA
43BSEQ-R	TGGTAGTAAACTTCGGTGAA
43PCR-R	AGACTCAAGCTGGTAGTAAACTTCG
44PCR-F	CTACCAAATTGGTGGTTATTCTGAG
44ASEQ-F	GGTGGTTATTCTGAGGATAG
44ASEQ-R	GATGGCTAGTGTGACTAGCA
44BSEQ-F	TATGTACTCATTCGTTTCGG
44BSEQ-R	AAATTGTAGTAACATAATCC
44PCR-R	TAGAATAGGCAAATTGTAGTAACAT
45PCR-F	ATTACCGTTGAGGAGCTTAAACAAC
45ASEQ-F	AGGAGCTTAAACAACTCCTG
45ASEQ-R	CAATGACAAGTTCACTTTCC
45BSEQ-F	TCAATGTGGTCATTCAACCC
45BSEQ-R	ATTGTAACCTGGAAGTCAAC
45PCR-R	TATCTCTGCTATTGTAACCTGGAAG
46PCR-F	ACAGACCACGCCGGTAGCAACGACA
46ASEQ-F	CCGGTAGCAACGACAATATT
46ASEQ-R	GGGTGAAATGGTGAATTGCC
46BSEQ-F	GCGAGCTATATCACTATCAG

(*continued*)

Table 4 (*Continued*)
Primer Sequences

46BSEQ-R	ATTAAAACAAGGAATAGCAG
46PCR-R	AATAAGCATTATTAAAACAAGGAAT
47PCR-F	TCACCATTAAGAGAAAGACAGAATG
47ASEQ-F	GAGAAAGACAGAATGAATGA
47ASEQ-R	GGATCTTGACAGTTGATAGT
47BSEQ-F	CTGCTTGGCTTTGTGCTCTA
47BSEQ-R	CTTGCCATGCTGAGTGAGAG
47PCR-R	TAAGTTCCTCCTTGCCATGCTGAGT
48PCR-F	CGCAATGGGGCAAGGCCAAAACAGC
48ASEQ-F	CAAGGCCAAAACAGCGCCGA
48ASEQ-R	TTCCTTGAGGAAGTTGTAGC
48BSEQ-F	TGGGTTGCAACTGAGGGAGC
48BSEQ-R	TTTTTGGCGAGGCTTTTTAG
48PCR-R	TGGCAGTACGTTTTTGGCGAGGCTT
49PCR-F	AGCAAAGTTTCTGGTAAAGGCCAAC
49ASEQ-F	CTGGTAAAGGCCAACAACAA
49ASEQ-R	GTGGGAATGTTTTGTATGCG
49BSEQ-F	ACTTATCATGGAGCCATTAA
49BSEQ-R	CTACTTGTGCTGTTTAGTTA
49PCR-R	TTAACTAAACCTACTTGTGCTGTTT
50PCR-F	TGGGCTATGTAAACGTTTTCGCAAT
50ASEQ-F	AAACGTTTTCGCAATTCCGT
50PCR-R	TTACACATTAGGGCTCTTCCATATAGG

Sequences for primary (PCR-F and PCR-R) and secondary (PCR-F and BSEQ-R) PCR primers. ASEQ-F, BSEQ-F, ASEQ-R, BSEQ-R are the sequencing primers. The sequences are from 5' to 3'.

3.7. Sequence Comparison

1. Edit, align, and compare sequences using the Tor2 strain (GenBank accession number AY274119) as a reference with the software designed for this purpose, for example SeqScape (*see* **Note 9**).
2. Re-sequence regions that reveal nucleotide substitutions using a combination of different primer sets to ensure the quality of the sequencing data (*see* **Note 10**).

4. Notes

1. For RNA extraction, carrier poly(A) RNA is added to the lysis buffer to increase the yield. Because the PCR primers are specific to the SARS-CoV genome, the subsequent amplification would not be affected. However, if one wants to perform 3' rapid amplification of cDNA ends (3' RACE) or similar cloning operation that depends on oligo(dT) priming of poly(A) tail of the viral RNA, then the carrier poly(A) RNA should be avoided.

2. Our previous studies have shown that, with two rounds of PCR, even direct clinical samples can be sequenced. This obviates the need for viral culture, which may pose a health hazard if not handled properly. It also minimizes the possible generation of viral mutants through culturing. However, direct clinical samples of high viral titer and a sensitive PCR amplification are required.

3. Yields of viral RNA should be determined by quantitative RT-PCR, because spectrophotometric determination is prone to error as a result of low RNA quantity and interference by the carrier poly(A) RNA, which contributes to most of the RNA present.

4. A prolonged denaturation step is used to remove secondary RNA structures in the SARS-CoV genome that impede reverse-transcription. The use of random hexamer ensures an even representation of the whole RNA genome and allows more sequence information to be obtained from a limited amount of viral RNA.

5. We recommend the use of a reverse transcriptase with increased thermal stability, which facilitates reverse-transcription at a higher temperature (55°C) than normal (42°C). This unfolds some of the secondary RNA structures, and thus produces longer cDNA at higher yields.

6. We recommend the simultaneous use of two different DNA polymerases in the PCR amplification. For example, the cDNA polymerase mix that we use contains Klen*Taq*-1 DNA polymerase, and a second DNA polymerase with 3' to 5' proofreading activity. The inclusion of a minor amount of a proofreading polymerase results in an error rate that is significantly lower than that for *Taq* alone *(18)*. This advantage is obvious when one is concerned about genomic sequence variations between different viral strains. The use of a two-polymerase system also increases the efficiency and yield, and hence the sensitivity, which is important when the viral titer is suboptimal.

7. The carryover of unused PCR primers into the sequencing reaction would lead to poor sequencing results. Like the sequencing primers, these unused PCR primers would also bind nonspecifically to the sequencing template in the cycle sequencing reaction, and, hence, generate noisy sequencing traces overshadowing the intended traces. Purification of the PCR products is, thus, usually recommended prior to their use as sequencing templates. However, these methods are labor-intensive and pose extra contamination risk, as they involve additional steps of opening and handling PCR products. Notably, we have suggested an optimized PCR protocol for direct sequencing of PCR products without PCR product purification. With the low PCR primer concentrations and the optimal number of cycles, most of the PCR primers are consumed at the end of the PCR. Furthermore, a nested sequencing primer selectively extends the specific PCR product in the cycle sequencing reaction. This would suppress any nonspecific PCR product from extension. The combined effect is a neat sequencing trace.

8. The amount of PCR product used for the sequencing reaction must be optimized carefully with different sequencing systems. Although more PCR product input usually gives higher signal intensities, it may also give shorter read lengths and oversaturated signals.

9. The PCR primers target 50 700-bp amplicons that overlap with each other along the SARS-CoV genome. The sequencing primers are designed in such a way that any sequence masked over by the PCR primer binding sites and the sequencing primer peak on one amplicon are reliably backed up by the homologous sequence in the overlapping amplicon.

10. We advocate scrutinizing efforts in validating any genomic sequence variation by resequencing regions with different combination of primers and sequencing chemistry. Because variation seen in a single viral isolate could potentially be a result of sequencing artifacts, we consider only the genomic sequence variations that are shared by at least two SARS-CoV isolates.

Acknowledgments

This work is supported by a Special Grant for SARS Research (CUHK 4508/03M) from the Research Grants Council of the Hong Kong Special Administrative Region (China).

References

1. Poutanen, S. M., Low, D. E., Henry, B., et al. (2003) Identification of severe acute respiratory syndrome in Canada. *N. Engl. J. Med.* **348,** 1995–2005.

2. Lee, N., Hui, D., Wu, A., et al. (2003) A major outbreak of severe acute respiratory syndrome in Hong Kong. *N. Engl. J. Med.* **348,** 1986–1994.

3. Tsang, K. W., Ho, P. L., Ooi, G. C., et al. (2003) A cluster of cases of severe acute respiratory syndrome in Hong Kong. *N. Engl. J. Med.* **348,** 1977–1985.

4. Rota, P. A., Oberste, M. S., Monroe, S. S., et al. (2003) Characterization of a novel coronavirus associated with severe acute respiratory syndrome. *Science* **300,** 1394–1399.

5. Marra, M. A., Jones, S. J., Astell, C. R., et al. (2003) The genome sequence of the SARS-associated coronavirus. *Science* **300,** 1399–1404.

6. Ksiazek, T. G., Erdman, D., Goldsmith, C. S., et al. (2003) A novel coronavirus associated with severe acute respiratory syndrome. *N. Engl. J. Med.* **348,** 1953–1966.

7. Drosten, C., Gunther, S., Preiser, W., et al. (2003) Identification of a novel coronavirus in patients with severe acute respiratory syndrome. *N. Engl. J. Med.* **348,** 1967–1976.

8. Peiris, J. S., Chu, C. M., Cheng, V. C., et al. (2003) Clinical progression and viral load in a community outbreak of coronavirus-associated SARS pneumonia: a prospective study. *Lancet* **361,** 1767–1772.

9. Ng, E. K. O., Hui, D. S. C., Chan, K. C. A., et al. (2003) Quantitative analysis and prognostic implication of SARS coronavirus RNA in the plasma and serum of patients with severe acute respiratory syndrome. *Clin. Chem.* **49,** 1976–1980.

10. Ng, E. K. O., Ng, P. C., Hon, K. L., et al. (2003) Serial analysis of the plasma concentration of SARS coronavirus RNA in pediatric patients with severe acute respiratory syndrome. *Clin. Chem.* **49,** 2085–2088.

11. Chim, S. S. C., Tsui, S. K. W., Chan, K. C. A, et al. (2003) Genomic characterisation of the severe acute respiratory syndrome coronavirus of Amoy Gardens outbreak in Hong Kong. *Lancet* **362,** 1807–1808.

12. Tsui, S. K. W., Chim, S. S. C., and Lo, Y. M. D. (2003) Coronavirus genomic-sequence variations and the epidemiology of the severe acute respiratory syndrome. *N. Engl. J. Med.* **349,** 187–188.

13. Ruan, Y. J., Wei, C. L., Ee, A. L., et al. (2003) Comparative full-length genome sequence analysis of 14 SARS coronavirus isolates and common mutations associated with putative origins of infection. *Lancet* **361,** 1779–1785.

14. He, J. F., Peng, G. W., Min, J., et al. The Chinese SARS molecular epidemiology consortium (2004) Molecular evolution of the SARS coronavirus during the course of the SARS epidemic in China. *Science* **303,** 1666–1669.

15. Chim, S. S. C., Tong, Y. K., Hung, E. C. W., Chiu, R. W. K., and Lo, Y. M. D. (2004) Genomic sequencing of a SARS coronavirus isolate that predated the Metropole Hotel case cluster in Hong Kong. *Clin. Chem.* **50,** 231–233.

16. Chiu, R. W. K., Chim, S. S. C., and Lo, Y. M. D. (2003) Molecular epidemiology of SARS—from Amoy Gardens to Taiwan. *N. Engl. J. Med.* **349,** 1875–1876.

17. Chiu, R. W. K., Chim, S. S. C., Tong, Y. K., et al. (2005) Tracing SARS-coronavirus variant with large genomic deletion. *Emerg. Infect. Dis.* **11,** 168–170.

18. Barnes, W. M. (1994) PCR amplification of up to 35-kb DNA with high fidelity and high yield from lambda bacteriophage templates. *Proc. Natl. Acad. Sci. USA* **91,** 2216–2220.

Index